The American Medical Association

HOME MEDICAL LIBRARY

GENES AND INHERITANCE

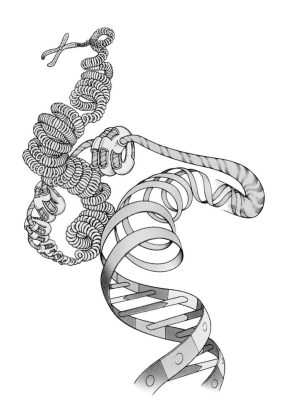

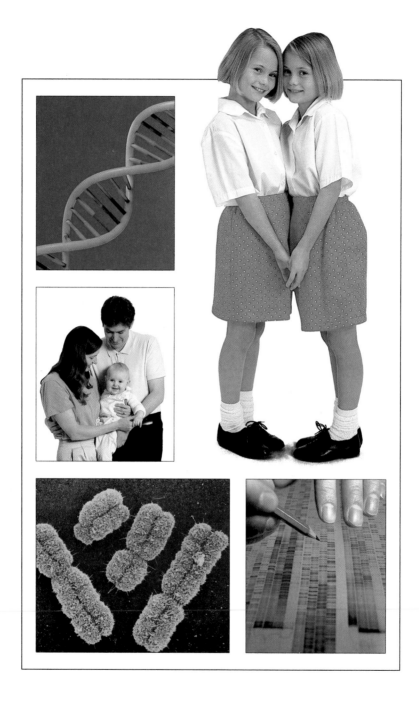

THE AMERICAN
MEDICAL ASSOCIATION

GENES AND INHERITANCE

Medical Editor
CHARLES B. CLAYMAN, MD

THE READER'S DIGEST ASSOCIATION, INC.
Pleasantville, New York/Montreal

The information in this book reflects current medical knowledge. The
recommendations and information are appropriate in most cases;
however, they are not a substitute for medical diagnosis. For specific
information concerning your personal medical condition, the AMA
suggests that you consult a physician.

The names of organizations, products, or alternative therapies appearing
in this book are given for informational purposes only. Their inclusion
does not imply AMA endorsement, nor does the omission of any
organization, product, or alternative therapy indicate AMA disapproval.

The AMA Home Medical Library is distinct from and unrelated to the
series of health books published by Random House, Inc., in conjunction
with the American Medical Association under the names "The AMA Home
Reference Library" and "The AMA Home Health Library."

Library of Congress Cataloging in Publication Data

Genes and inheritance / medical editor, Charles B. Clayman.
       p.   cm. — (The American Medical Association home medical
library)
    At head of title: The American Medical Association.
    Includes index.
    ISBN 0-89577-460-7
    1. Medical genetics—Popular works.   2. Human genetics—Popular
works.   I. Clayman, Charles B.   II. American Medical Association.
III. Series.
RB155.G3587 1993
616'.042—dc20   92-12671

FOREWORD

Nothing promises to change the field of medicine as much as the science of genetics. We are on the verge of developing new medical treatments and cures that only recently were not thought possible. In fact, until the middle of this century, little was known about our genes and how they work. Now we are learning that these units of hereditary information cause or contribute to our most common disorders, including cancer and heart disease, as well as to many rare conditions. In what promises to open up medicine's newest and potentially most fruitful frontier, scientists are learning how to supplement or replace disease-causing genes with healthy ones to cure even the most devastating inherited disorders.

This volume of the AMA Home Medical Library explains the genetic revolution in detailed yet understandable language. You will learn how DNA – the chemical substance that genes are made of – was discovered and the crucial role that genes play in our lives and health. In fact, your identity as an individual is a direct result of the interaction of your genes with the environment. Your genes largely determine your characteristics, your susceptibility or resistance to diseases, your body's reaction to medications, whether you carry and can transmit a genetic disease to your children, or whether you have or will develop an inherited disorder yourself.

In this volume, we explain how genes can become altered to cause diseases of all kinds, and we describe the sophisticated new laboratory tests that can detect many hereditary disorders even before birth. We hope this volume will help you and your family understand how doctors can predict your risk of developing a genetic disease and how doctors can, in many cases, help you diminish those risks.

James S. Todd MD

JAMES S. TODD, MD
Executive Vice President
American Medical Association

CONTENTS

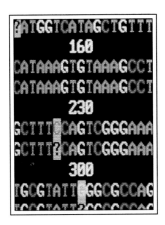

CHAPTER ONE

WHAT ARE GENES?

I T IS OFTEN EASY to see that children resemble their parents – not only in appearance, but in personality traits and even susceptibility to certain diseases. But it has only been in the last 150 years that science has uncovered the molecular basis for these observations. Applying this knowledge to the understanding and treatment of illness began only recently.

The first part of this chapter traces the path that scientists took toward understanding heredity. Long before they had the technology to examine the detailed makeup of cells, scientists suspected that what people inherit from their parents is controlled by specific combinations of chemicals that direct biological processes. When they watched cells divide under a microscope and saw threadlike structures called chromosomes,

they reasoned that the chromosomes were somehow involved in transmitting characteristics from one generation to the next. Because they observed an abundance of human characteristics and relatively few chromosomes, researchers deduced that each chromosome contained distinct hereditary units responsible for those characteristics. They called those units "genes."

A major scientific breakthrough in understanding heredity was the discovery of the structure of the principal chemical compound that makes up chromosomes – deoxyribonucleic acid (DNA). DNA is present in all living cells and carries all the information required for life. You might call it an instruction manual for life processes. In humans, this instruction manual exists in the form of 46 "chapters," or chromosomes, half of which are contributed by each parent.

Genes can be thought of as sentences that each carry an instruction for a specific characteristic (such as eye color) or function (such as hormone production).

Although it is our genes that make us unique, the similarities among us are greater than the differences. Because all forms of life have evolved over billions of generations from common ancestors, many of our genes are identical to those of other animals and even to plants. Chimpanzees share most of our genes, cows share many, carrots and cabbages share a few. The difference lies in how the genes are put to work.

The influence of genes on some traits, such as intelligence and behavior, is difficult to determine. Your identity as an individual results from the complex interaction of your genes and your environment.

A SHORT HISTORY OF GENETICS

Farmers and animal breeders have long relied on heredity to improve the quality of plants and animals by crossbreeding. Until recently, however, people had little idea how the complex mechanisms of heredity worked. The science of genetics is now providing answers. Genetics is being used to solve medical mysteries and to diagnose and treat illness. In the near future doctors hope to use the new knowledge to cure and even to prevent many common genetic disorders.

THEORIES OF EVOLUTION

For centuries, people believed that all species of plants and animals were created at about the same time. However, fossil studies have shown that most species on Earth evolved over hundreds of millions of years; that is, their structure changed gradually over generations. These gradual alterations are responsible for the astonishing diversity and countless number of species alive today, including our own. Since the early 19th century, scientists have debated how these changes occurred.

Fossil studies
Fossil studies continue to provide us with information about the plants and animals that have lived on Earth over hundreds of millions of years. The fish at left, called a coelacanth, was known only from fossils and thought to be extinct until living descendants (below) with slightly altered characteristics were found earlier this century.

Homunculus
At the time of the early Greeks, many people believed that each sperm contained a miniature person called a homunculus that entered a woman's uterus and grew there until it was born. It was not until the middle of the 19th century that studies of embryos disproved this idea and scientists performed experiments that revolutionized our understanding of heredity.

Acquired characteristics

In the early 19th century, French biologist Jean Baptiste Lamarck (1744-1829) introduced the idea that plants and animals changed or adapted as they reacted to changes or new threats in their environment. Lamarck theorized that the characteristics that an organism altered or acquired during its life were transmitted to offspring. Widely accepted at the time, Lamarck's ideas were later discounted in favor of Darwin's theory of evolution by natural selection.

THE GALÁPAGOS FINCHES

While visiting the Galápagos Islands off South America, Darwin observed that different species of finches on different islands resembled a single species of finch on the South American mainland. Darwin thought it likely that the finches had colonized the islands from the mainland many years before and had then been isolated. He later postulated that a variety of species evolved over many generations on the different islands as the birds adapted to distinctly different environments.

Natural selection

In 1859, British naturalist Charles Darwin (1809-1882), in his book *The Origin of Species,* presented his theory of the way in which species originated and evolved through generations. From his careful observation of plants and animals, Darwin realized that the members of any species vary a little in their characteristics and that, because of this variation, some members are better adapted than others to their environment. This led him to suggest that gradual changes occur in a species over generations because better adapted members survive longer and produce more offspring than those who are less well adapted. In Darwin's theory of "survival of the fittest," those organisms that are "fit" (better adapted) would be "naturally selected" over organisms with less helpful traits. They would then pass those advantageous traits on to offspring, who would also be more likely to survive and reproduce and pass on the same beneficial traits. In this way, the traits that help a species adapt to its environment would gradually become more common. Conversely, the traits that reduce an organism's chances for survival and reproduction would gradually become rare within a species over generations.

THE LAWS OF HEREDITY

Long before scientists understood the molecular basis of heredity, a part-time scientist established some of the basic rules. From his studies on the garden pea, Gregor Mendel (1822-1884), an Austrian monk and teacher, formulated an accurate account of how genetic characteristics are transmitted from one generation to the next (see page 12). His work led to the new field of genetics and even today provides fresh insights into the important role that genes play in our lives.

Darwin's unifying theory
In his theory of evolution, Charles Darwin (left) offered the first plausible scientific explanation of how life – in its awesome variety and complexity – evolved on this planet. He provided a unifying concept into which all subsequent discoveries in genetics and other fields of biology could be integrated.

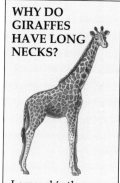

WHY DO GIRAFFES HAVE LONG NECKS?

Lamarck's theory suggested that giraffes developed long necks because they had to stretch to feed on the leaves of trees; the acquired characteristic of a long neck was then passed on to offspring. According to Darwin's theory, those ancestors of today's giraffes that were born with longer necks were better equipped to eat the leaves off the treetops and, therefore, better able to survive and reproduce. Their offspring were more likely to have long necks and, in turn, survive and reproduce.

Beak variation
Darwin found that the size and shape of the beaks of different finches were related to their diets. The medium ground finch (top row left), which feeds mainly on large seeds, has a short, heavy beak. The cactus ground finch (bottom row right) has a long beak designed for probing flowers of the prickly pear, on which it feeds.

MENDEL: THE FIRST GENETICIST

Gregor Mendel (left), like his contemporary Charles Darwin, realized that species of plants and animals differed considerably in characteristics such as size, shape, and color. Mendel decided to find out what determined a species' traits and how those traits were inherited. He chose the garden pea as his research subject. Mendel studied seven of the plant's traits, each of which had two distinct forms – the stems, for example, were either tall or short.

Seven pairs of characteristics

Mendel studied pea plants to determine how certain characteristics are inherited. Of the seven traits that Mendel studied, those he called dominant are shown in the left column below; the corresponding characteristics that he called recessive are in the right column.

Round seeds   **Wrinkled seeds**

Yellow seeds   **Green seeds**

Purple flowers   **White flowers**

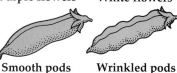

Smooth pods   **Wrinkled pods**

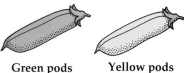

Green pods   **Yellow pods**

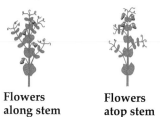

Flowers along stem   **Flowers atop stem**

Tall stems   **Short stems**

1 In a process called purebreeding, Mendel established strains of pea plants for each trait that he wanted to study. For example, he established a strain of tall plants, and a strain of short plants. Mendel then crossbred plants from these pure strains. He crossed a tall plant with a short plant (right), and a plant that produced round seeds with a plant that produced wrinkled seeds (below right).

2 Mendel found that offspring (called hybrids) from these crossbreedings never exhibited a blend of characteristics. Instead, they displayed just one of the traits. For example, all the offspring of the crosses between tall and short plants were tall; all the offspring of the crosses between plants with round and wrinkled seeds had round seeds. Mendel called the persistent traits, such as tall stems and round seeds, "dominant." The forms that seemed to have disappeared, such as short stems and wrinkled seeds, he called "recessive."

3 Mendel then took the hybrid plants produced from his first generation of crossbreedings and crossed them with each other. The offspring of this second generation of crosses (right) followed two remarkable patterns. First, some of the offspring exhibited the recessive trait (such as a short stem), which had not appeared in the first generation. Second, for every one offspring with the recessive trait (a short stem), there were three plants with the dominant trait (a tall stem).

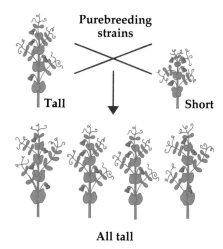

Purebreeding strains

Tall   Short

All tall

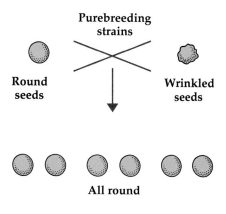

Purebreeding strains

Round seeds   Wrinkled seeds

All round

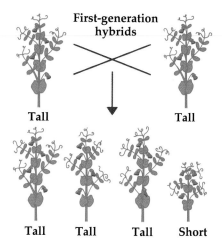

First-generation hybrids

Tall   Tall

Tall   Tall   Tall   Short

4 Mendel's experiments are illustrated below with the heredity of tall and short stems. He theorized that each of the traits he studied was determined by the interaction of a pair of factors (later called genes) that can come in one of two forms. Mendel referred to the alternate forms of the factors that determine stem length as T (for long stems) and t (for short stems). He determined that, if both forms are present in a plant, one form (T) is dominant and it masks the effect of the other form (t), which is recessive.

5 The purebred tall plants had the factors TT; purebred short plants had the factors tt. The offspring of the first round of crosses were all Tt but, because T was dominant, the plants all had long stems. Each plant passes on only one of its two factors to each offspring. When the first-generation hybrids were crossed with each other, an average of one in four offspring inherited the recessive (t) factor from both parent plants and, therefore, had short stems. The others inherited at least one dominant (T) factor, giving them long stems.

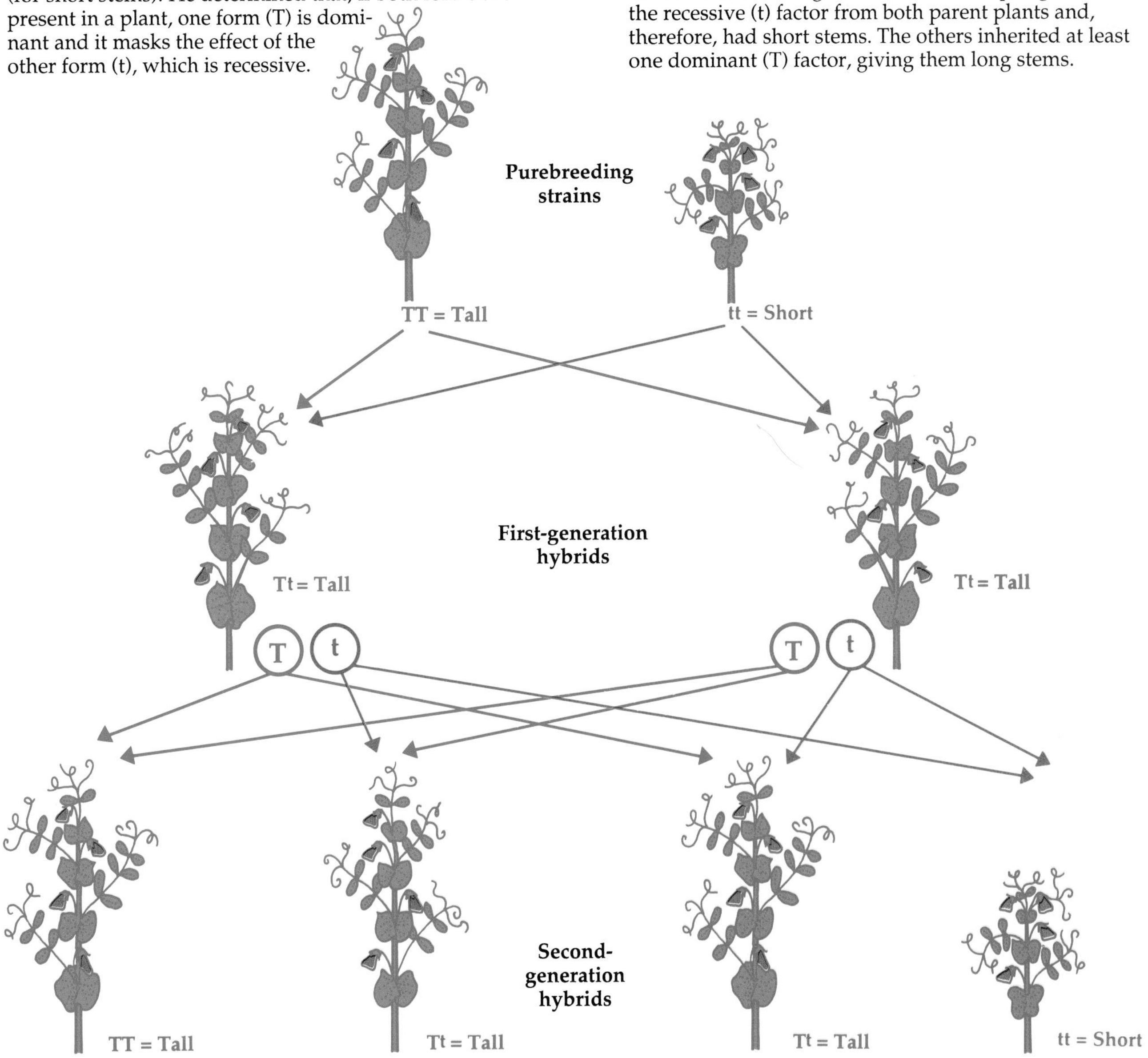

Purebreeding strains

TT = Tall   tt = Short

First-generation hybrids

Tt = Tall   Tt = Tall

T   t   T   t

Second-generation hybrids

TT = Tall   Tt = Tall   Tt = Tall   tt = Short

WHY WERE MENDEL'S PEAS IMPORTANT?

Through his experiments, Mendel established the following fundamental principles of heredity:

◆ The inherited traits of a living organism are determined by distinct factors (now called genes) that are transmitted from one generation to the next.

◆ Many characteristics of organisms are determined by the interaction of a pair of hereditary factors.

◆ The factors in a pair determining a trait may be the same or different. When they are different, the dominant form masks the effect of the other form, which is called recessive.

◆ Organisms with specific traits are produced in predictable numbers according to statistical rules from one generation to the next.

Mendel's rules form the foundation of the science of genetics and our understanding of how traits are inherited.

THE DISCOVERY OF CHROMOSOMES

In 1882, German anatomist Walther Flemming (1843-1905) was experimenting with synthetic dyes that would enable him to examine transparent animal cells under a microscope. He noticed that some parts of the cell absorbed the dyes while other parts did not. The dyes were absorbed most by a defined area inside the cell called the nucleus. Flemming named the stained material "chromatin," from the Greek word for color. He observed that the structure of the chromatin changed when cells divided to form two identical daughter cells. At the beginning of the process of cell division, the threadlike material condensed into a number of short, compact segments. Flemming's colleague Heinrich Wilhelm Waldeyer (1836-1921) later named these compact segments chromosomes, a term meaning "colored bodies."

Cells and chromosomes
In 1882, after Walther Flemming first observed chromosomes inside a cell (below, magnified 1,500 times), Edouard van Beneden, a Belgian cell expert, showed that each cell in animals of the same species contained the same number of chromosomes. He also found that sperm and egg cells contained exactly half the number of chromosomes present in other cells.

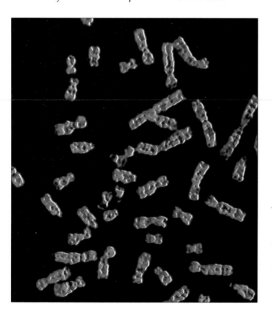

Purple eyes	Red eyes	Brown eyes	White eyes

DISCOVERING THE GENE

After the discovery of chromosomes, many scientists suspected that chromosomes were the carriers of the hereditary factors that Gregor Mendel had studied several decades before in his experiments with peas. However, it was unclear how the 40 to 50 chromosomes then estimated to be contained in human cells could account for the far greater number of human characteristics. We now know that people have 46 chromosomes – half from each parent – arranged in 23 pairs. By the early 1900s scientists suggested that each chromosome carries many distinct hereditary units that determine the characteristics of a plant or animal. They called those units genes. At the time, no one had any idea of what a gene actually looked like. Nevertheless, the notion that a gene is the hereditary factor responsible for a single characteristic or function became widely accepted.

Chromosomes and DNA
Early geneticists understood that chromosomes contained genetic material, but knew little about how this material was arranged. We now know that chromosomes are threadlike strands of DNA (right, magnified 50,000 times). Just before cells begin to divide to make copies of themselves, the individual strands of DNA become tightly coiled into the compact X shape of chromosomes (above right, magnified 10,000 times).

Chromosomes and genes
In the 1930s, zoologist Thomas Morgan used fruit flies in his genetics experiments because they breed rapidly. By studying the heredity of a variety of characteristics such as eye color, Morgan proved that chromosomes are the carriers of inherited factors and that these factors – genes – are arranged in linear form along the threadlike chromosomes.

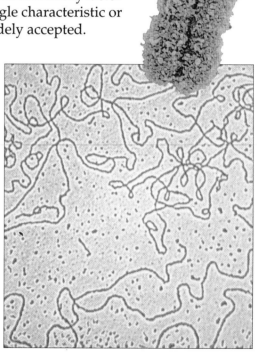

THE SIGNIFICANCE OF DNA

For many years, scientists believed that protein was the genetic material inside chromosomes because proteins seemed to be the only substances complex enough to serve as storehouses of information for life-forms. But, in 1944, American scientist Oswald Avery published a paper describing his experiments with bacteria that challenged that notion and revolutionized the understanding of heredity up to that time. Avery's experiments showed that when deoxyribonucleic acid (DNA), a substance present in the chromosomes of all living cells, was transferred from one bacterium to another, it radically changed the characteristics of the recipient bacterium.

In 1952, two American scientists, Alfred Hershey and Martha Chase, proved that DNA is indeed the storehouse of genetic information. In experiments using viruses made only of DNA and protein, they showed that viruses reproduce inside infected bacteria even when their protein has been removed. Hershey and Chase concluded that only DNA is required for replication and, therefore, is the genetic material that determines the traits of an organism.

A model for DNA
Watson and Crick (left) built a wire model to test whether their theory of DNA's spiral structure complied with the rules of chemistry. Everything fit perfectly. Their results, published in 1953, sparked an explosion of research that changed the direction of biology. They solved the mystery of how the genetic code maintains its structure and is replicated and read.

THE STRUCTURE OF DNA

British chemist Francis Crick and American geneticist James Watson hypothesized that the DNA molecule formed a double spiral. Chemical analysis confirmed that the DNA molecule contains long chains that wind around each other, resembling a twisted ladder – the double helix. The chains, or sides of the ladder, are made of structural sugars and phosphates. They are connected by pairs of chemical compounds, called nucleotide bases, that form the rungs. The four bases are adenine, cytosine, guanine, and thymine (shown below).

The double helix
Watson and Crick saw that the number of units of the chemical base adenine in the DNA molecule was about the same as the number of units of the base thymine and that the number of units of guanine was about the same as the number of units of cytosine. They proposed that the bases occur in pairs – in which adenine links only to thymine and cytosine links only to guanine – to form a ladder between a double spiral backbone of sugar and phosphate. The double helix structure keeps the genetic code in sequential order and enables DNA to replicate when cells divide.

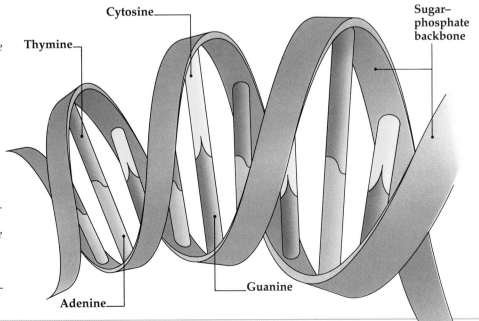

Cytosine

Thymine

Sugar–phosphate backbone

Adenine

Guanine

ANALYZING CHROMOSOMES

In 1969, Down's syndrome became the first disorder to be detected by analyzing the chromosomes of a fetus before birth. Since then, more than 600 different chromosome abnormalities have been identified. In the US, about 20,000 babies are born each year with a chromosome defect. All the recognized chromosome abnormalities can be diagnosed by examining cells under a high-power microscope. Only a small number of the disorders caused by single genes can be diagnosed in this way.

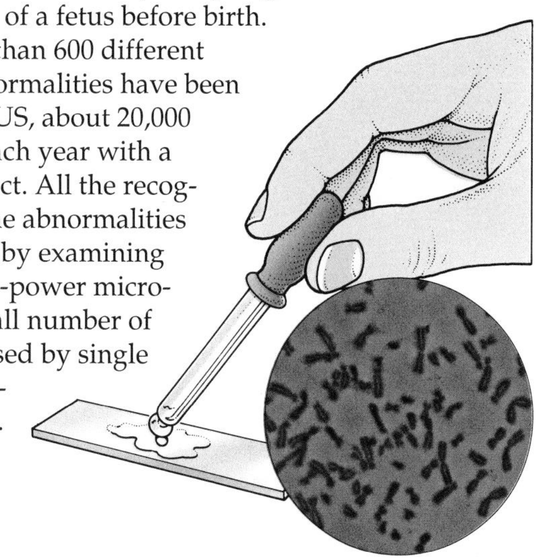

MANIPULATING DNA

The deciphering of the genetic code in the 1960s accelerated genetic research. The first synthetic gene was produced in 1970 and many techniques for manipulating DNA have since been developed. A major advance was the discovery of restriction enzymes – biological "scissors" that can cut DNA at specific sites – which enable scientists to identify and study individual genes. Another breakthrough was the development of genetic probes. These short lengths of DNA of a known base sequence are radioactively labeled for easy detection when they link up with their matching base sequence on an unidentified DNA strand. Scientists are using these new biological tools to discover and locate genes.

THE GENETIC CODE

The discovery of the structure of DNA led to the realization that genes consist of long sequences of chemicals called nucleotide bases whose order forms a kind of code. Scientists found that the lengths of DNA that make up genes instruct chemical compounds (amino acids) inside cells to string together in a particular way to form proteins. In 1961, Francis Crick and colleague Sidney Brenner proved that the genetic code works in triplets; that is, each sequence of three bases is the code for a specific amino acid. The same year, American biochemist Marshall Nirenberg determined the three bases that provide the code for the amino acid phenylalanine. By the mid 1960s, the base triplet codes for all 20 amino acids had been deciphered. Crick arranged the codes into a meaningful order that eventually became the standard genetic code dictionary. The dictionary is the reference for interpreting the genetic code of all living organisms. Once a gene's code is known, scientists can use that information to make copies of the gene in the laboratory.

Genetic engineering
In genetic engineering, a gene that makes a specific protein is transferred from one organism to another, enabling a recipient organism to make the protein in large quantities. In the example shown here, the gene for human insulin is obtained from human DNA, spliced to a piece of DNA from a bacterium, and then put back into the bacterium. The bacterium multiplies rapidly, producing an abundance of human insulin used for treating diabetes.

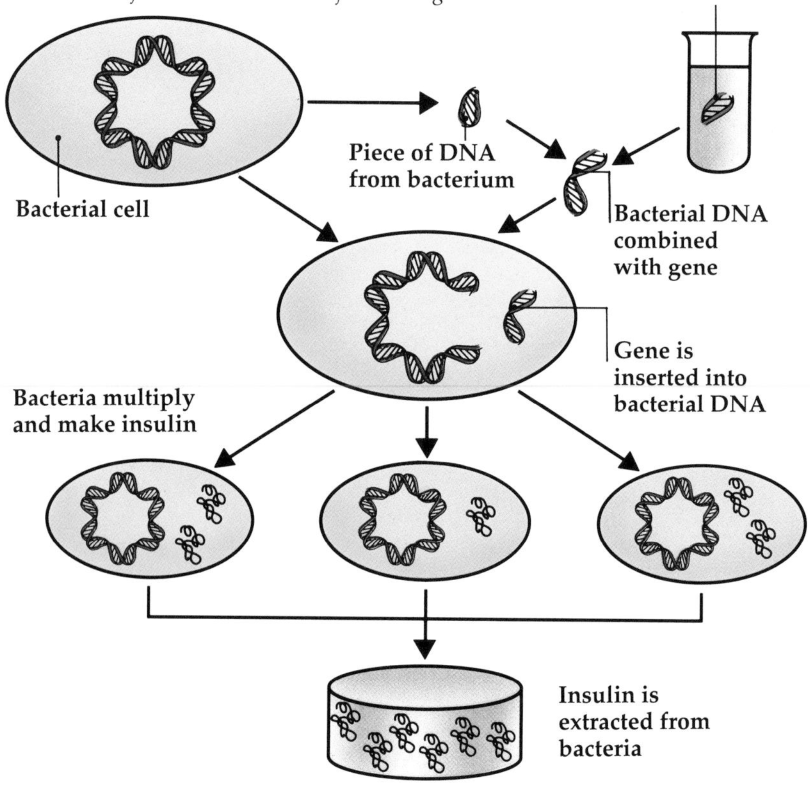

Gene that makes human insulin
Bacterial cell
Piece of DNA from bacterium
Bacterial DNA combined with gene
Gene is inserted into bacterial DNA
Bacteria multiply and make insulin
Insulin is extracted from bacteria

THE HUMAN GENOME PROJECT

When it became possible to understand the genetic instructions contained inside our cells, biologists dreamed of some day reading the complete text written in the language of DNA. A massive international scientific effort, called the human genome project, plans to do just that. The project is comparable in complexity and expense to putting a person on the moon. Scientists estimate that listing the full complement of human DNA instructions from beginning to end would fill 13 sets of encyclopedias.

By 1992, more than 2,000 of our approximately 100,000 genes had been identified. Genes are already being used to produce human proteins in large quantities to treat diseases. One genetically engineered protein – human growth hormone – is helping many children who are deficient in the hormone to grow normally. However, huge commercial and ethical implications arise from the isolation of a gene. Many scientists feel they are entitled to a patent for a gene they discover that might be used commercially. An international committee has been established to develop a consensus on the social, ethical, and legal issues raised by the genome project.

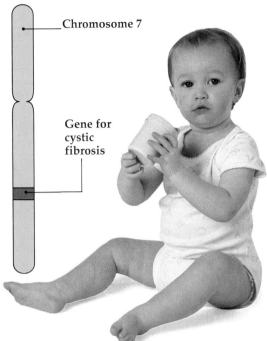

Chromosome 7

Gene for cystic fibrosis

Cystic fibrosis gene
More than 1,000 children are born each year in the US with cystic fibrosis, a fatal disorder affecting the lungs. In 1989, scientists located the cystic fibrosis gene on chromosome 7 (far left). This scientific breakthrough made prenatal diagnosis and carrier testing possible and paved the way for new treatments and even a cure.

Analyzing the genome
Scientists throughout the world are involved in mapping the human genome – the complete set of human genes. This awesome project is expected to take 15 years and to cost 3 billion dollars. However, much more time will be required to fully understand the genetic code and its role in development, health, and illness.

DNA FINGERPRINTS

Every person's DNA has a unique pattern referred to as his or her DNA fingerprint. DNA fingerprints are used increasingly to identify with high probability the biological father of a child and to link criminals to biological material, such as hair, semen, or skin scrapings, found at a crime scene. In the criminal identification at right, the DNA pattern from suspect 1 does not match that of the crime sample. The pattern from suspect 2 is identical to the crime sample, thus linking suspect 2 to the crime.

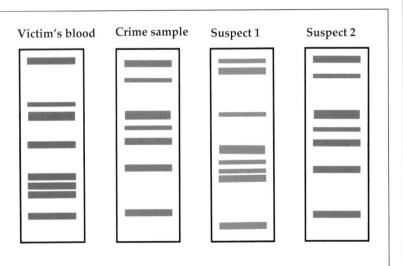

Victim's blood   Crime sample   Suspect 1   Suspect 2

GENES AND ENVIRONMENT

Each of us is born with a set of genes that plays a role in determining what we will look like and how our bodies will develop and function. But the way we look and function is also influenced by everything that happens to us in the uterus, and in infancy, childhood, and adult life. These environmental influences, which begin during fetal development, include exposure to infections and to poisons such as alcohol and chemicals from cigarette smoke and pollution. Other factors that can affect the function of genes are diet, climate, exercise, body chemistry, and accidents or infectious diseases. Although cells have an amazing ability to repair genes damaged by such environmental factors, some defective genes manage to escape detection and cause disease. Ultimately, what happens to our bodies is the result of a complex interplay of heredity and environment.

GENETICALLY DETERMINED TRAITS

Genes control some of your characteristics, such as your blood type and eye color, that cannot be influenced by your environment. Some disorders are caused by defective genes that a person is born with. Other disorders, such as cancer, may be caused by genes that become damaged by environmental factors such as tobacco smoke and sunlight.

Hair
The color and texture of your hair are determined by genes but can be altered to some extent by environmental influences such as sunlight and diet.

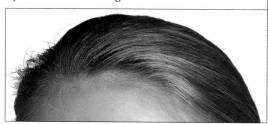

FACIAL APPEARANCE: NATURE OR NURTURE?

Your facial features are determined largely by the set of genes you inherited from your parents. That unique set of genes is called your genotype. However, your appearance can also be influenced by environmental factors such as diet and disease. The way in which you function and appear at any given time as the result of a combination of genetic and environmental influences is called your phenotype.

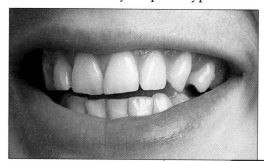

Teeth
The size and shape of your teeth are determined by your genes. But environmental factors such as diet, poor dental hygiene, disease, or injury can cause significant changes in the appearance of teeth.

Nose
The size and shape of your nose are genetically determined. Only injury or cosmetic surgery can change its appearance. However, alcohol, drugs, and even drinking too many caffeine-containing beverages can influence the color and texture of the overlying skin.

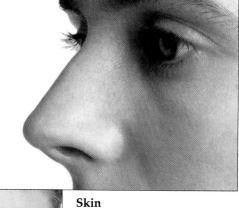

Skin
Genes determine the color and texture of your skin, but exposure to the sun can make it darker, wrinkled, and leathery. Your skin's texture can also be influenced by other factors, including diet and smoking.

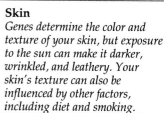

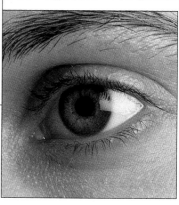

Eyes
The color of your eyes is determined by your genes. Environmental influences cannot change brown eyes to blue eyes, although factors such as aging and poor health may cause slight changes.

HEIGHT POTENTIAL

As with most variable traits, height is the result of a subtle interplay of genes and environment. Your genes determine your potential maximum height, but environmental factors – especially your diet – can influence your growth. A poor diet can significantly stunt growth and prevent a person from reaching his or her genetically determined maximum height.

Diet and stature
The children of Japanese people who migrated to Hawaii grew to be an average of 2 to 4 inches taller than their parents. The height differences between the generations have been linked to a change from a low-protein diet to a high-protein diet.

Nearsightedness
Nearsightedness, the inability to see distant objects clearly, is an inherited trait. However, not everyone who inherits the gene or genes for nearsightedness is affected.

ENVIRONMENTALLY INFLUENCED TRAITS

Some of your traits and abilities are determined by your environment. For example, the language or languages you are taught as a child depend on the culture in which you are raised. Although the ability to use language well might run in families, this trait is heavily influenced by environmental factors such as growing up in a home in which reading is a frequent activity. Many of our likes and dislikes – such as preferences in art, music, and even foods – are also determined primarily by the influences of our culture and upbringing.

CONTINUOUSLY VARIABLE TRAITS

Many traits that are known to be influenced by both genes and environment are called continuously variable traits. They are usually complex physiological or behavioral traits, such as weight and blood pressure, that can vary in response to many factors and can be measured.

Because relatives share some of the same genes, it is tempting to attribute their similarities to genetic factors. But members of the same family also tend to share the same environment, making it difficult to disentangle the effects of genes from those of environment.

HANDEDNESS

The brain is divided into two hemispheres. Some brain functions, such as handedness, are located in either the left or the right hemisphere. About 90 percent of us use our right hand for writing and two thirds of us favor our right hand for other activities requiring coordination and skill. The remaining 10 percent are left-handed or ambidextrous (able to use both hands equally well). Because the world is set up to accommodate right-handed people, parents of left-handed children sometimes encourage them to use their right hand. But handedness cannot be changed easily because it is mainly an inherited trait – left-handed people are more likely to have left-handed children.

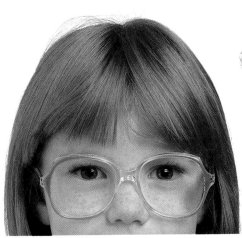

Causes of mental retardation

Mental retardation can be caused by abnormalities in a person's genes or chromosomes or by environmental factors such as diseases or injuries that affect the brain. Many causes of abnormal brain development are unknown.

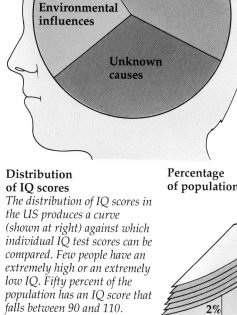

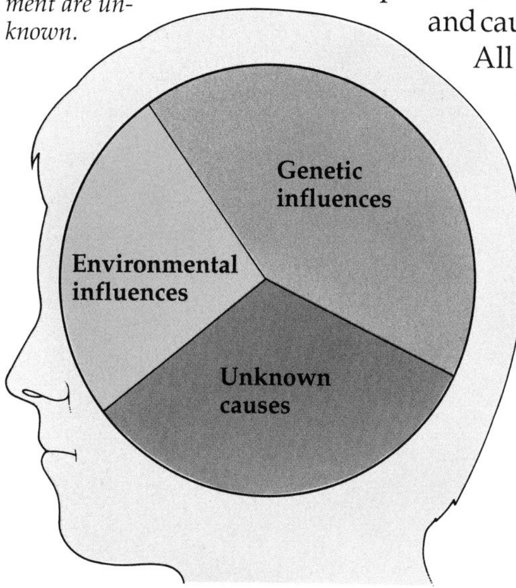

INTELLIGENCE

Genes undoubtedly influence our intelligence. Some genetic abnormalities can lead to severe mental retardation. For example, in an inherited metabolic disorder called phenylketonuria (PKU), toxins (poisons) accumulate in the brain and cause mental retardation. All newborns are tested for PKU. Children with the disorder are given a special low-protein diet. This restricted diet – if started in the first 2 months of life and followed throughout life – allows children with the disorder to develop normally. PKU and its treatment demonstrate how genes and environment together can influence intelligence.

Measuring intelligence

Studies of twins reared apart provide evidence of the important role that genes play in determining intelligence. Intelligence continues to be evaluated by IQ (intelligence quotient) tests. Although mental abilities are too complex to be accurately measured, IQ tests are often used to predict success in school or in a certain occupation.

While IQ tests can be a useful arbitrary measure of intelligence of people in the same culture, they may indicate little more than a person's ability to perform well on IQ tests. No test can reliably compare, for example, the intelligence of an American college student with the intelligence of a person of the same age living in the Amazonian rain forest.

Distribution of IQ scores

The distribution of IQ scores in the US produces a curve (shown at right) against which individual IQ test scores can be compared. Few people have an extremely high or an extremely low IQ. Fifty percent of the population has an IQ score that falls between 90 and 110.

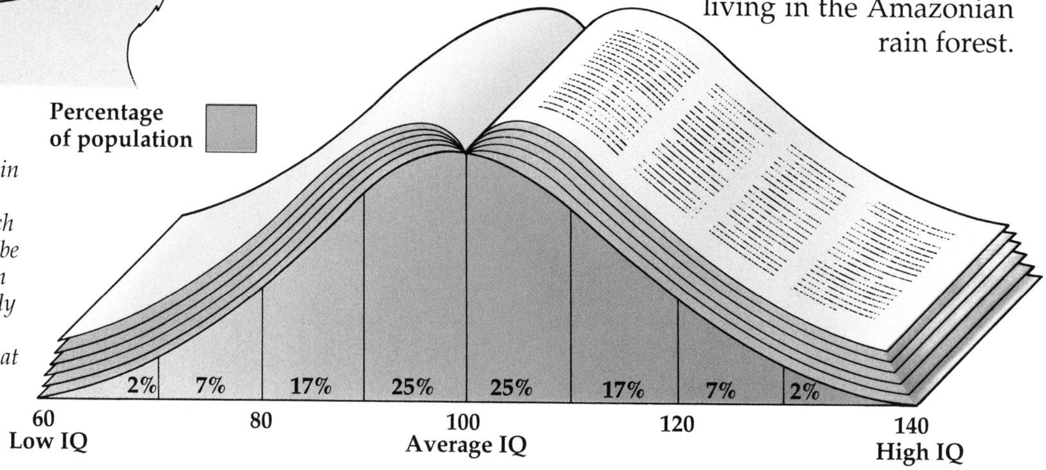

Percentage of population

| 60 Low IQ | 2% | 7% | 17% | 25% | 25% | 17% | 7% | 2% | 140 High IQ |

80   100 Average IQ   120

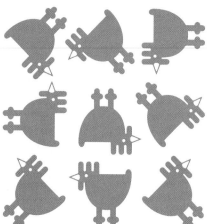

IQ tests

IQ tests vary in the way they attempt to measure intelligence. Many tests have a large number of questions that require sophisticated language skills to answer correctly. Two examples of nonverbal IQ test questions are shown here (the answers can be found on page 144). Nonverbal IQ tests are helpful in assessing the intelligence of people from diverse cultural backgrounds because the tests do not rely on language skills.

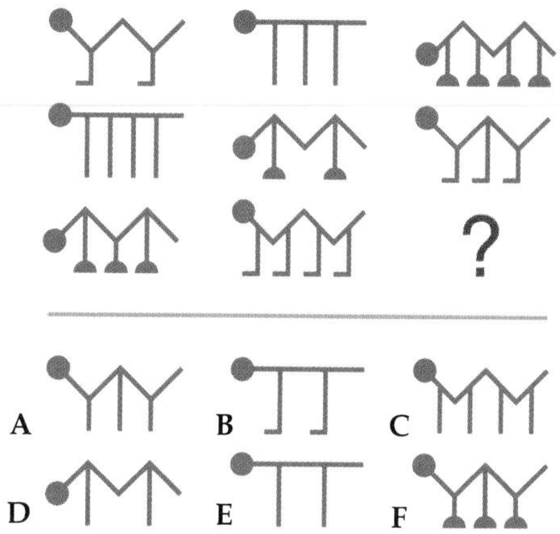

Which chicken does not match the others?

Which figure (A to F) completes the pattern?

BEHAVIOR

Much of your behavior is instinctive; that is, you respond to a particular stimulus in a predictable way. These responses, or reflexes, are programmed by your genes. Like other animals, humans have genes that influence their behavior to help them to survive and reproduce. These beneficial genes are the driving force of evolution.

Genes do not control your behavior directly. You rely on your brain and nervous system to help you respond to life from day to day. But genes may be responsible for programming your nervous system with "response guidelines," including basic survival responses such as jumping out of the way of a car speeding toward you. You are also programmed with a capacity for learning and the ability to modify your behavior as a result of your experience.

The extent to which human behavior is genetically programmed is unknown. Some people believe that behavior is ultimately determined by genes – a view that leaves little room for free will. A more widely accepted theory is that we are able, to some extent, to overrule the influence of our genes.

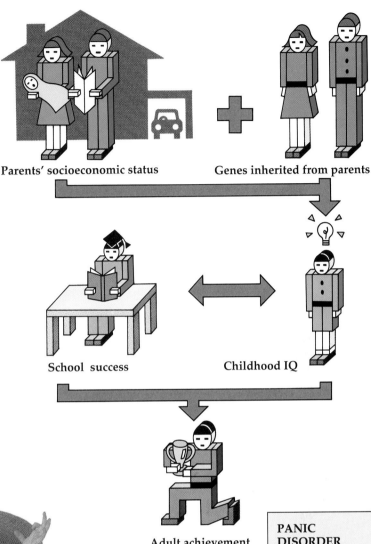

Parents' socioeconomic status   Genes inherited from parents

School success   Childhood IQ

Adult achievement

Learned behavior
Children learn how to behave by watching and imitating their parents. Acceptable behavior is also learned through messages conveyed in speech. Much of our behavior can therefore be considered to be acquired through cultural influences rather than through the biological influence of our genes.

IQ and success
Many factors interact to determine your intelligence and eventual success in life. These influences include the genetic components of intelligence and the environment in which you are raised. Studies that measured socioeconomic status by parents' education and occupation, number of siblings, and quantity of educational materials in the home found that people of higher socioeconomic status typically scored higher on IQ tests.

PANIC DISORDER

Recent research suggests that there is a genetic component in panic disorder. The condition, which involves sudden and repeated attacks of overwhelming anxiety, occurs in only 1 percent of the general population. However, it occurs in about 15 percent of people from families in which one person is affected. Researchers think that one dominant gene is responsible.

TWIN STUDIES

"Heritability" is the extent to which genetic factors account for the variation in a trait. A trait such as eye color, which is primarily influenced by genes, has a high heritability. Conversely, a trait such as intelligence has a lower heritability because it is also influenced by many environmental factors such as education.

Comparing identical and nonidentical twins

Studies of twins have found that some traits, such as height and intelligence, correlate much more closely in identical twins (who are genetically identical) than in nonidentical twins (who are no more alike genetically than any other two siblings). These findings indicate that many of our traits are determined largely by our genes. The heritability of height is estimated to be about 94 percent, that of intelligence to be roughly 50 percent. However, because identical twins tend to be treated alike, the environmental influences on them may enhance their similarities in a number of traits.

Identical twins
Although identical twins have the same genes, they may not look exactly alike or behave in the same way. The differences are determined by their responses to the influences of their environment.

Nonidentical twins
In a pair of nonidentical twins, each one has a unique set of genes. Nonidentical twins are no more alike genetically than any two siblings.

Identical twins raised apart
While identical twins who were raised by different parents (right) share many similarities in behavior and appearance, they often have some differences too.

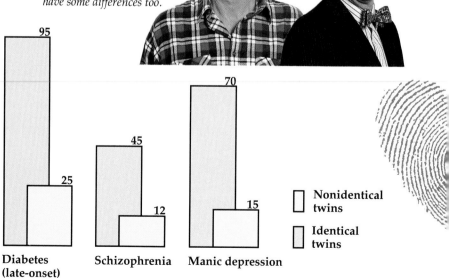

Disease susceptibility
The chart below shows the degree of heritability of certain disorders, based on studies of twins raised apart. Twins are said to be concordant if they both have the same disorder. Disorders that are shared more often by identical twins than nonidentical twins are more likely to be hereditary. Late-onset diabetes has a strong genetic influence; cancer has a strong environmental influence.

Concordance (percent)

100 —

Allergies	Cancer	Diabetes (late-onset)	Schizophrenia	Manic depression
50 / 4	17 / 11	95 / 25	45 / 12	70 / 15

☐ Nonidentical twins

☐ Identical twins

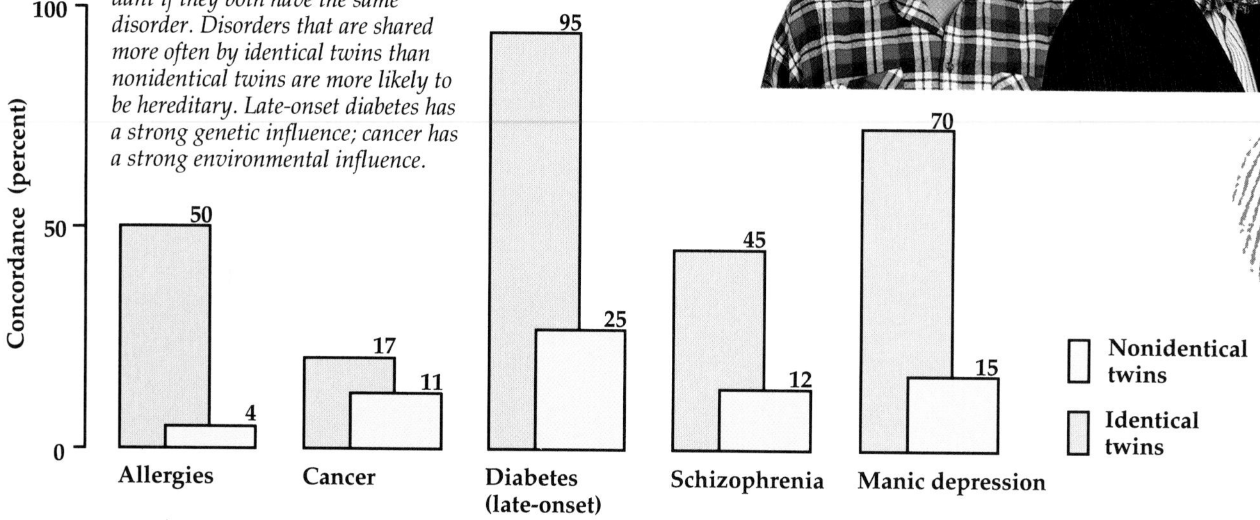

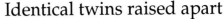

Sleep patterns
Studies show that the sleep patterns of identical twins are more similar than those of nonidentical twins, suggesting a genetic influence. The studies examined daytime napping, regular bed time, and sleep duration.

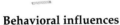

Identical twins raised apart
Studies of identical twins who were raised apart have helped to determine the influence of genes on complex traits such as intelligence. A strong similarity in a trait in identical twins raised apart indicates a strong genetic influence because their genes are identical but their environments are different.

These studies suggest that genes play an important role in determining intelligence. However, the validity of some of the studies has been questioned in cases in which the twins were raised in similar environments.

A Swedish study of 650 pairs of twins found that identical twins weighed about the same as adults whether or not they had been raised together. Nonidentical twins did not have this similarity. Twin studies also indicate that alcohol addiction and some mental disorders are influenced by genes.

Behavioral influences
Many parents – consciously or unconsciously – encourage identical twins to dress alike and participate in the same activities. These influences may affect the results of studies comparing identical and nonidentical twins.

IQ studies
Studies comparing the IQs of identical twins have found a 90 percent rate of similarity. The IQs of nonidentical twins were similar only 50 percent of the time. These findings suggest that genetic factors play a significant role in determining intelligence.

Fingerprints
Everyone has a unique fingerprint that does not change after birth. Even identical twins have unique, although very similar, fingerprints, suggesting that subtle variations in the environment inside the uterus may affect the development of fingerprint pattern.

Twins and obesity
When different pairs of identical twins were given the same diet, the twins in a pair tended to gain similar amounts of weight. However, among the different pairs, the amount of weight gain varied, indicating that the tendency to put on weight may be genetically influenced.

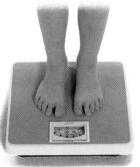

CHAPTER TWO

HOW GENES WORK

THE UNION OF A SPERM and an egg at the moment of conception produces a fertilized egg that contains, in the form of the chemical DNA, all the instructions necessary to make a person. DNA continuously makes copies of itself and directs the production of other chemicals inside cells. Under the control of DNA, the fertilized egg divides many times, allowing the embryo to develop and grow. In a healthy fetus, DNA ensures that specialized cells, such as liver and brain cells, are produced in the right place, at the right time, and in precisely the right amount. DNA gives the cells their specialized functions. For example, DNA instructions provide nerve cells with receptors that allow the cells to send and receive chemical and electrical messages. The first sections of this chapter examine the way in which DNA's special structure enables it to regulate all the activities inside your cells. DNA controls cells through instructions located along its double strands in the form of a code. Short sections of the DNA strands, called genes, act as individual units of information. Their job is to direct the production of chemicals called proteins. Some proteins are used by the body to form organs and tissues such as skin and muscles. Other proteins, called

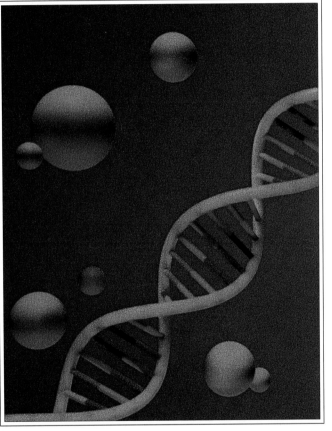

antibodies, are used to fight infection; enzymes and some hormones are proteins that act as controlling influences in the body's chemical processing.

Most cells inside the body continually divide to provide for its growth and repair. With each division, a cell's genetic material is faithfully and efficiently copied to newly formed cells. However, egg and sperm cells are produced differently, by means of a special type of cell division that involves a mixing of genetic material. This random mixing ensures that each person (except for identical twins) has a unique combination of genes. The two types of cell division, and the significance of errors that can occur during these processes, are discussed in the sections on gene transmission and genetic mutation and variability. Changes in DNA ultimately account for the genetic variation that exists among us.

Because half of your genes come from each of your parents, you are likely to resemble your close relatives. The final section in this chapter highlights inheritance patterns and the way in which our genes influence our appearance and identity. Some traits are determined by just one pair of genes. Others are determined by the combined effect of many genes and environmental factors.

GENES: BODY BUILDERS

Each of us begins life as a single cell containing genetic information inherited from our mother and father. That cell divides to form two identical cells, and the process of division continues until, as adults, we have about 100 trillion cells. Each of our cells contains an identical set of genes that is an exact copy of the set of genes contained in the original cell.

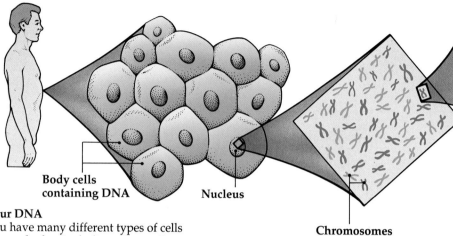

Centromere

Genes

Chromosome

Body cells containing DNA

Nucleus

Chromosomes

Your DNA
You have many different types of cells in your body – including nerve cells, muscle cells, liver cells, and fat cells – but each cell contains exactly the same genetic material (DNA) inside its nucleus. Cells differ from each other because different genes are active in different cells. You might think of your DNA as an instruction manual (below), or blueprint, that makes you the person you are. You have 100 trillion identical copies of this instruction manual, one in every cell; each DNA manual holds all the information needed to construct, maintain, and repair your body.

Your chromosomes
The DNA inside the nucleus of each cell is packaged in 46 distinct segments called chromosomes (above), which can be thought of as 23 paired chapters of an instruction manual (below). Half of your chromosomes (shown here in red) came from your mother; the other half (shown here in blue) came from your father. Each chromosome consists of a long thread of DNA. Chromosomes appear as separate X-shaped segments just before cell division, when each has made a duplicate copy of itself. The two copies of a duplicated chromosome are held together at a constricted region called a centromere.

Your genes
Genes are distinct units along the threads of DNA that make up chromosomes. If a chromosome is a chapter of your DNA instruction manual, your genes are the sentences in the chapter (below). Just as each sentence might contain a specific instruction, each gene contains an instruction for some aspect of your body's development and functioning. There are about 100,000 genes in each cell nucleus, distributed among the 46 chromosomes. Each chromosome may contain a thousand or more genes.

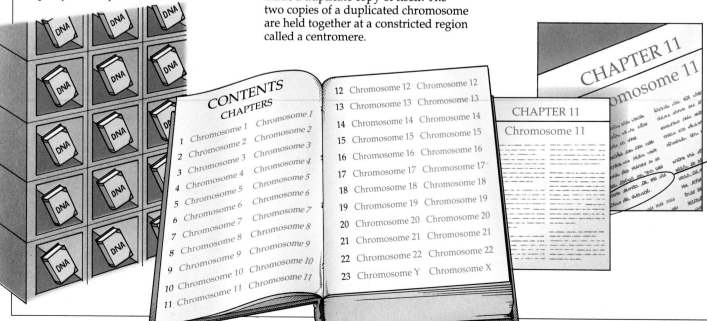

CONTENTS
CHAPTERS

CHAPTER 11
Chromosome 11

CHAPTER 11
Chromosome 11

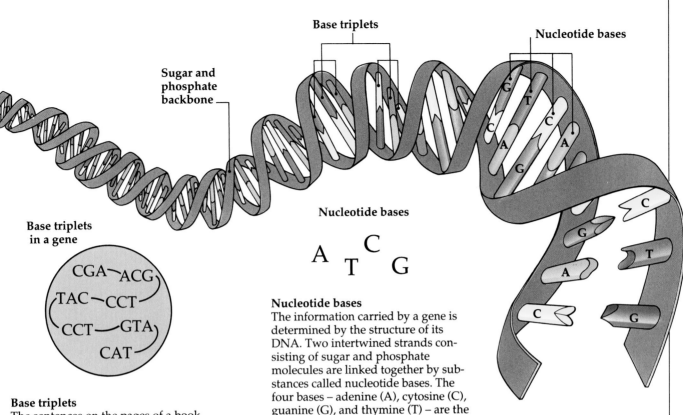

Base triplets

Sugar and phosphate backbone

Nucleotide bases

Nucleotide bases

Base triplets in a gene

CGA ACG
TAC CCT
CCT GTA
CAT

Nucleotide bases

A T C G

Nucleotide bases

The information carried by a gene is determined by the structure of its DNA. Two intertwined strands consisting of sugar and phosphate molecules are linked together by substances called nucleotide bases. The four bases – adenine (A), cytosine (C), guanine (G), and thymine (T) – are the letters of the words (base triplets) in a sentence (gene). AGT, TAC, CGG, and TCA are examples of such words, which are the codes for particular amino acids – the building blocks of proteins. By reading the sequence of base triplets in a gene and following those instructions, a cell can string together amino acids in the correct order to make a specific protein.

Base triplets

The sentences on the pages of a book consist of words that, when strung together, mean something to the reader (below). Similarly, genes also consist of words that must be read by body cells in order for the cells to follow the instructions for making specific enzymes and other proteins. The words of the genetic code are triplets of nucleotide bases called base triplets (above). Base triplets can be read and understood by using a standard genetic code dictionary that is common to all living organisms.

Letters in words

n d s w
c f b t
r i a o

Words in a sentence

String together
the following amino acids
to make the protein
insulin

HOW GENES CONTROL YOUR BODY

Genes direct the formation and functioning of all the organs and processes in your body. Throughout your life, your genes maintain and repair your body. They do so by regulating the production of many types of proteins essential for life. Scientists think that some base sequences in the DNA code serve as messengers to control genes; that is, they give instructions to genes such as "make more protein under this condition," "stop," or "slow down."

SUMMARY

GENETIC MATERIAL (DNA)	Instruction manual for life processes
CHROMOSOMES	Chapters
GENES	Sentences
BASE TRIPLETS	Words
NUCLEOTIDE BASES	Letters

GENES AND CHROMOSOMES

YOU INHERITED ALL of your genes from your parents – half from your mother and half from your father. The egg and the sperm from which you originated each contained exactly half of the genetic material found in all the cells of your body. Their union at the moment of conception produced a single cell containing the full set of genes necessary to make a new human being. All of your cells grew from that single fertilized egg cell.

The genes you inherited from your parents determine your physical characteristics, such as blood type and eye color. Genes also contribute to variable traits such as height and blood pressure, which can also be influenced by environmental factors such as diet. Your inherited characteristics depend on which genes from your parents were dealt to you at random at the moment of conception and the way in which those two sets of genes have interacted.

GENES AND CELLS

Although each of your cells (except for the egg or the sperm) contains an identical set of genes, specialized cells – such as muscle cells and nerve cells – have distinct functions. To allow a cell to perform its specified task, only a fraction of the genes are switched on in that cell. The genes that are activated enable the cell to do its job. The other genes in the cell are switched off. Some genes that provide for basic functions such as repair of cell walls are active in all cells. Genes control the production of proteins, which give cells their physical structure and allow them to carry out their chemical functions. The selective activation and deactivation of genes allow a cell to make only those proteins it is programmed to make.

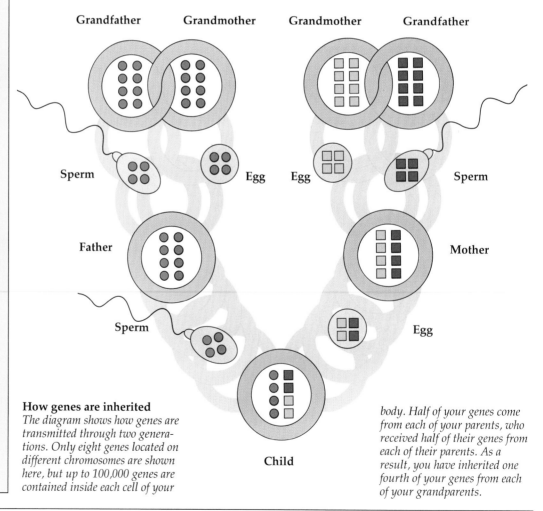

How genes are inherited
The diagram shows how genes are transmitted through two generations. Only eight genes located on different chromosomes are shown here, but up to 100,000 genes are contained inside each cell of your body. Half of your genes come from each of your parents, who received half of their genes from each of their parents. As a result, you have inherited one fourth of your genes from each of your grandparents.

Why are siblings different?

Each sperm that a man makes contains only half of his genes in a combination different from that in any other sperm. In the same way, each egg that a woman forms contains a different selection of half of her genes. For this reason, the children produced from the union of different pairs of eggs and sperm differ genetically from each other as well as from their parents. At right, letters represent genes.

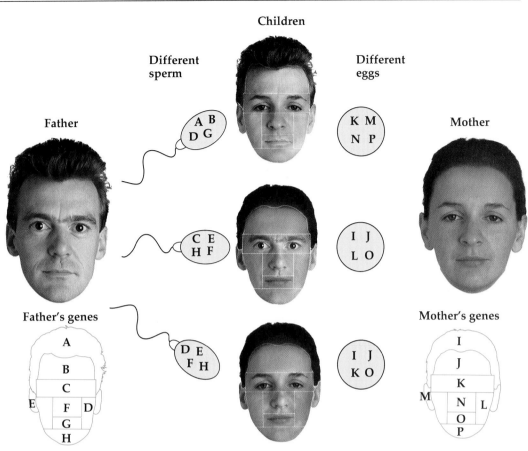

Children

Different sperm

Different eggs

Father

Mother

Father's genes

Mother's genes

CHROMOSOMES

Chromosomes usually cannot be seen inside a cell, even under a powerful microscope. However, when the cell begins to divide, the genetic material inside its nucleus contracts. If the cell is stained with a dye at this time, the shortened and thickened chromosomes can be seen under a microscope as dark rods. At this stage, each chromosome has just duplicated itself (see page 37). The two duplicate copies are joined together at a point called the centromere, which gives the chromosome an X shape.

All human cells, except the egg and the sperm, contain 46 chromosomes arranged in 23 pairs. In the union of an egg and a sperm (which each contain 23 chromosomes, exactly half the number found in other cells), each parent provides one member of all the 23 pairs of chromosomes needed to make a new person. Of the 23 pairs, 22 are the same in both males and females. These identical 22 chromosome pairs are called autosomes. The 23rd pair consists of the sex chromosomes. In women, these are two X chromosomes, which look alike but differ slightly in their genetic instructions. Men have two different sex chromosomes – the female X chromosome and the much shorter Y chromosome, which confers maleness.

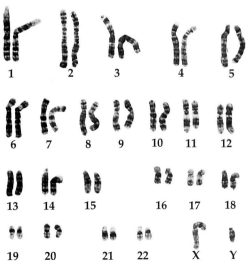

Your chromosomes

A karyotype is a picture of your chromosomes made by staining and photographing them and arranging them in pairs in a standard format. The microscopic "snapshot" is taken just before cell division, when chromosomes are most visible. In the male karyotype at left, the last pair of chromosomes consists of the X and Y sex chromosomes. In females, the last pair consists of two X chromosomes.

Although the chromosomes in the nucleus of each of your cells are made up of equal amounts of genetic material from your mother and father, there is some genetic material elsewhere in each cell that comes only from your mother. This genetic material is contained in structures called mitochondria, which are responsible for supplying a cell with energy. Mitochondria are located in the cell's cytoplasm (the area surrounding the nucleus), and they contain their own set of genes. Mitochondrial DNA functions independently of the DNA inside the nucleus. Mitochondria are present in large numbers in eggs but are absent from sperm. Geneticists attempting to trace the roots of humans living today believe that all of us have mitochondrial genes that have been passed down through many generations from one woman who lived about 200,000 years ago.

Structure of chromosomes

If the DNA contained in all the chromosomes in a cell were unpacked and strung out, it would form a line more than 6 feet long. DNA is tightly packaged, wound around special proteins that give it density. By weight, each human chromosome is made up of roughly equal amounts of DNA and a variety of proteins. A chromosome's rigid framework is made of structural proteins, and its outer surface is coated with different kinds of proteins that regulate chemical processes inside cells.

The amount of DNA in a cell is not necessarily an indication of the amount of genetic information that it carries. It is estimated that as much as 95 percent of our DNA does not control protein production or have any other known function. Some scientists believe that this so-called junk DNA may have, at some time in our evolutionary past, consisted of active, functioning genes.

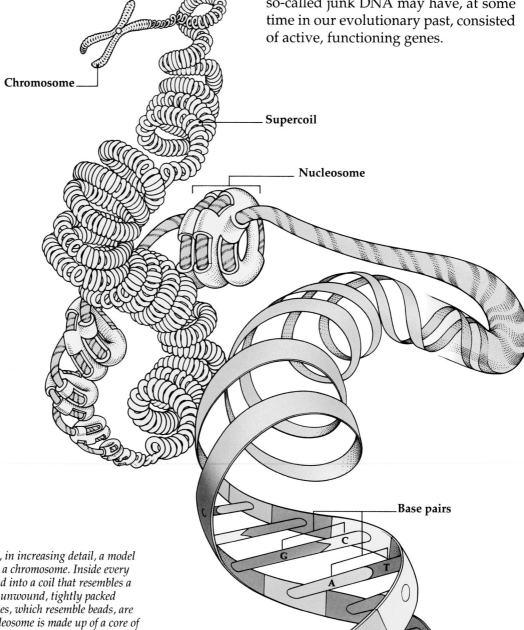

Chromosome

Supercoil

Nucleosome

Base pairs

DNA double helix

Coiling of DNA
The diagram at right shows, in increasing detail, a model of the packaging of DNA in a chromosome. Inside every chromosome, DNA is wound into a coil that resembles a telephone cord. If the coil is unwound, tightly packed structures called nucleosomes, which resemble beads, are visible in the coil. Each nucleosome is made up of a core of special binding proteins around which the DNA is wrapped several times. Further unwinding reveals the characteristic double-helix structure of the DNA molecule.

DOUBLE-HELIX STRUCTURE OF DNA

The most important property of DNA, provided by its unique structure, is its ability to replicate easily and accurately. DNA information is written out in a code along only one of its two strands (shown here in blue). The letters of the DNA code are four nucleotide bases. Because a nucleotide base on one strand will pair up only with a specific base on the other strand (shown here in green), the order of bases on the information strand exactly determines the order of bases on the other strand. Base A pairs up only with base T and base G pairs up only with base C. In this way, DNA faithfully makes a copy of itself.

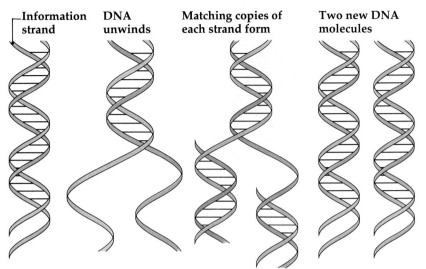

Information strand   **DNA unwinds**   **Matching copies of each strand form**   **Two new DNA molecules**

Replication of DNA
To copy itself, the DNA double helix unwinds and separates its two strands. In the process, new matching strands form on each of the two original strands, producing two identical copies of the entire DNA molecule.

Mapping human chromosomes

Generally, every gene has a 50-50 chance of being transmitted from a parent to a child. But some genes tend to be passed on in combinations; that is, they occur together more often than would be expected by chance. Because whole sections of chromosomes may be passed on when eggs and sperm are formed, genes located near each other on the same chromosome are likely to "travel" together. The closer that two genes are on a chromosome, the greater the chance that they will be transmitted together from parent to child. Once scientists identify a gene as a member of a particular group, they can map the location of other genes in the group. The ability to map or locate genes in relation to one another has proven invaluable in identifying many disease-causing genes, including the gene for cystic fibrosis. Such genetic linkage also enables doctors to track a specific DNA pattern and the heredity of a certain disorder in a family, to predict whether or not a child may have inherited the disease-causing gene.

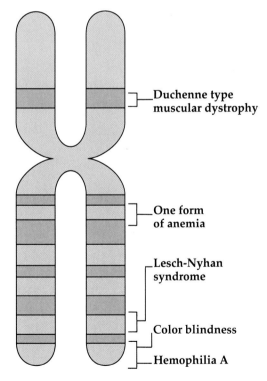

Duchenne type muscular dystrophy

One form of anemia

Lesch-Nyhan syndrome

Color blindness

Hemophilia A

Mapping the X chromosome
About one third of the more than 2,000 genes that have been identified so far are located on the X chromosome. The locations of some of them are shown above. X-linked genes are easier to identify because the disorders that they cause occur mainly in males. The blue and pink areas represent the chromosome's unique genetic banding pattern.

ONCOGENES

By-products of chemical reactions inside cells and environmental factors such as tobacco smoke and pollution can damage the genes that regulate normal cell division, turning them into cancer-causing genes called oncogenes. If the genes that promote cell division are damaged, they act like gas pedals that are stuck, causing the uncontrolled cell division that leads to the growth of a tumor. Other genes normally act like brakes to keep cell division in check. If these genes become damaged, cancer cells divide uncontrollably, destroying healthy tissue.

GENES AT WORK

GENES ARE RESPONSIBLE for the development, functioning, repair, and maintenance of all living organisms, including people. They do so by directing the production of essential chemicals called proteins. Each gene in every cell of your body provides instructions for making one protein or part of a protein.

Your body makes many different types of proteins. Most visibly, these proteins make up physical structures such as muscles and skin. In the form of enzymes and hormones, they control chemical processes in your body. Proteins called antibodies are an important part of your immune defense system.

GENE PRODUCTS

Proteins consist of chains of chemicals called amino acids. All human proteins are assembled from the same set of 20 amino acids in different numbers and combinations. A gene, or section of DNA, controls the production of a protein inside a cell by indicating which amino

DNA

RNA

Genetic information
DNA carries genetic information in a pattern of nucleotide bases that runs along its length. The nucleotide bases adenine (A), cytosine (C), guanine (G), and thymine (T) correspond to the four bases adenine, cytosine, guanine, and uracil (U) found on the RNA molecule. The bases are "read" by the cell in groups of three, called base triplets, each of which represents the code for an amino acid.

THE SEX GENE

In 1991, scientists located the gene that determines whether an embryo will develop into a male or a female. The gene was identified from a strand of DNA taken from the Y sex chromosome, which is found only in males. Scientists believe that the gene directs production of a protein that tells other genes to make still more proteins – the male sex hormones. These hormones influence cells to form male sex organs.

acids should be strung together and in what order. DNA controls the formation of proteins by sending out a "sidekick" called messenger RNA (ribonucleic acid) with the DNA's protein-building instructions. These coded instructions, or messages, carried by RNA are working copies of sequences of nucleotide bases found on sections of DNA (see below left). Messenger RNA carries the DNA instructions for making a specific protein (such as insulin) out of the nucleus into the area of the cell where proteins are made.

Controlling gene activity

In your body, you have many different types of cells with distinct shapes and functions. For example, brain cells differ significantly from muscle cells. Yet all of your cells are derived from a single fertilized egg and they all contain identical genes. Cells differ because different combinations of genes are switched on and off in each cell, determining which proteins the cell makes and enabling it to perform its specialized task.

The control of genetic activity inside cells is enormously complex. Hormones and enzymes produced by "control" genes regulate the production of proteins inside cells. Protein production is also directed by signals in the DNA molecule itself. All of these regulatory mechanisms work together in a wondrously orchestrated fashion to switch genes on and off or instruct them to go fast or slow. During fetal development, this regulatory activity directs the formation of all tissues and organs.

PROTEINS: THE FABRIC OF LIFE

Proteins are essential for human development and growth. They are an important part of the structures of your body, from your skin to your blood cells, and they carry out many of your body's vital chemical functions. Some proteins form structures such as muscles and hair. Other proteins – including enzymes, antibodies, and many hormones – dissolve in your body's fluids and are carried to wherever they are needed. Your genes control the manufacture of proteins in precisely the right number and at precisely the right time.

Many hormones are proteins
Many hormones, such as human growth hormone and insulin, are proteins that control a variety of biological activities. Hormones control growth, cell metabolism, sexual development, and specialized activities in some organs. The hormone prolactin, which is produced by the pituitary gland, stimulates milk production after a woman has given birth.

Prolactin stimulates milk production

Proteins help muscles work
Muscles are made of bundles of specialized cells that can contract and relax to create movement. Each cell in a muscle is made up of many different proteins. Muscles give you mobility, pump blood throughout your body, and help move food through your digestive tract.

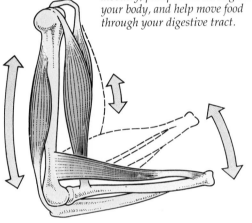

Antibodies are proteins
Proteins called antibodies are produced by white blood cells in the immune system to destroy disease-causing organisms, such as bacteria and viruses. Antibodies formed in response to vaccinations provide protection against specific infections.

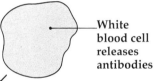

White blood cell releases antibodies

Antibodies

Antibodies attack invading organism

Organism

Another type of white blood cell engulfs and destroys organism

Target chemicals

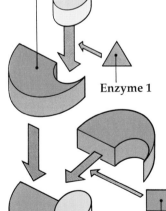

Enzyme 1

Enzymes are proteins
Enzymes are proteins that regulate the rate of chemical reactions in cells. Each enzyme stimulates a change in a specific chemical. An enzyme may split a chemical into separate pieces, join chemicals together, or simply modify a chemical's structure slightly. Enzymes work in sequence to promote chemical changes (see right) that convert raw materials into a finished product. For example, enzymes work together in liver cells to convert glucose (the body's fuel) into fat for energy storage.

Enzyme 2

Product

Hemoglobin is a protein
Hemoglobin is a protein that carries oxygen in red blood cells and gives the blood cells their red color. One red blood cell contains an average of 350 million hemoglobin molecules.

Oxygen-carrying site

Coiled strands of protein

HOW ARE PROTEINS MADE?

Protein production begins when DNA inside a cell nucleus "unzips" and the genetic code on one of its strands is converted into a matching code on a strand of RNA, called messenger RNA. This process is called transcription. Messenger RNA carries the genetic instructions out of the nucleus to a structure inside the cell called a ribosome. There, the coded message is translated into a sequence of amino acids. Other molecules in the cell – transfer RNAs – pick up and deliver amino acids to the ribosome "work-bench" where they attach to the corresponding amino acid codes on messenger RNA. The amino acids string together to make pro-teins. This highly choreographed process (called translation) takes place in your cells millions of times every second.

2 RNA nucleotide bases are paired up with their matching DNA bases and joined to form a strand of messenger RNA. The copying of the DNA strand ends at a sequence of bases (the termination code) that stops the transcription process.

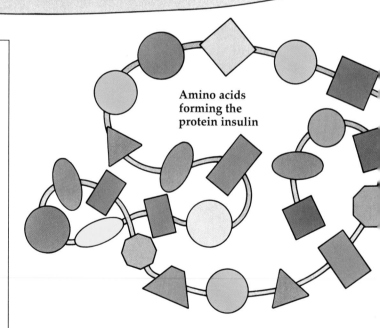

DNA nucleotide bases

RNA nucleotide bases

Termination code

1 The transcription process, in which DNA produces a single-strand replica called messenger RNA, begins when a particular enzyme attaches to a specific site on the DNA chain. The enzyme causes a section of the DNA helix to split open, forming two strands.

DNA

TRANSLATION OF THE GENETIC CODE

Like DNA, the RNA molecule carries four different nucleotide bases in various groups of three along its length. These nucleotide bases, adenine (A), cytosine (C), guanine (G), and uracil (U), carry the instructions for making a protein. Each triplet of nucleotide bases is the code for an amino acid. The sequence of these codes on the RNA strand determines the sequence of amino acids to make a particular protein in a process called translation (step 6 at right).

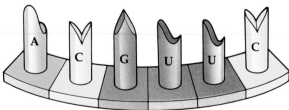

Base triplets
The sequence of base triplets (each representing an amino acid) in messenger RNA is read and translated into a series of amino acids to make a specific protein. Protein production, or translation, takes place on structures (called ribosomes) inside cells. Although each base triplet represents the code for one amino acid, different triplet codes can represent the same amino acid.

Amino acids forming the protein insulin

7 A termination signal at the end of the strand of messenger RNA stops the translation process and the completed chain of amino acids is released. The amino acid chain undergoes further chemical processing that converts it into an active protein (in this case, insulin).

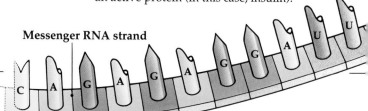

Messenger RNA strand

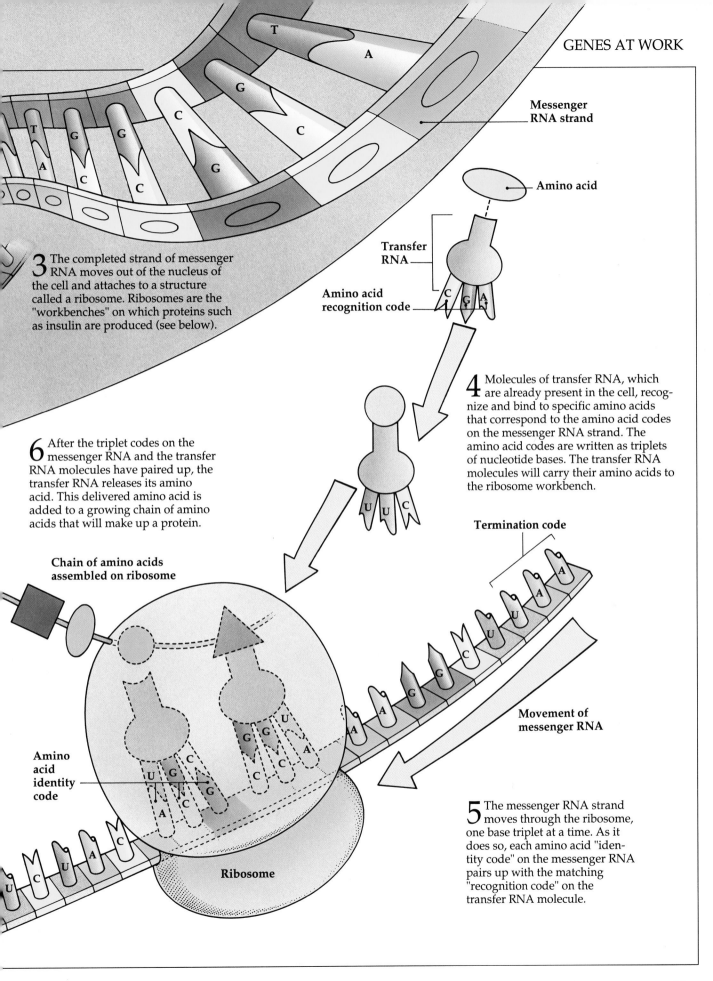

Messenger RNA strand

Amino acid

Transfer RNA

Amino acid recognition code

3 The completed strand of messenger RNA moves out of the nucleus of the cell and attaches to a structure called a ribosome. Ribosomes are the "workbenches" on which proteins such as insulin are produced (see below).

4 Molecules of transfer RNA, which are already present in the cell, recognize and bind to specific amino acids that correspond to the amino acid codes on the messenger RNA strand. The amino acid codes are written as triplets of nucleotide bases. The transfer RNA molecules will carry their amino acids to the ribosome workbench.

6 After the triplet codes on the messenger RNA and the transfer RNA molecules have paired up, the transfer RNA releases its amino acid. This delivered amino acid is added to a growing chain of amino acids that will make up a protein.

Termination code

Chain of amino acids assembled on ribosome

Movement of messenger RNA

Amino acid identity code

Ribosome

5 The messenger RNA strand moves through the ribosome, one base triplet at a time. As it does so, each amino acid "identity code" on the messenger RNA pairs up with the matching "recognition code" on the transfer RNA molecule.

35

HOW GENES ARE TRANSMITTED

GENES ARE CONTINUOUSLY reproduced in your cells as part of a simple copying process called mitosis. However, egg and sperm cells, through which genes are transmitted from parents to child, are produced by a more complex form of cell division called meiosis. The knowledge of how cells divide and how DNA copies itself has helped scientists to uncover the mechanisms of heredity and the origin of many genetic disorders.

Mitosis is a simple process of cell division that enables all many-celled living organisms to grow and to repair and maintain themselves. In each mitotic cell division, the genetic material (DNA) inside a cell is copied and distributed to two identical new cells. Mitosis occurs in your body thousands of times every second as dead or dying cells are replaced and new cells are formed. The process is shown at right under MITOSIS: CELL MULTIPLICATION BY DIVISION.

DNA REPLICATION

Before a cell can divide, the DNA inside its nucleus must make a copy of itself. To do so, the double strands of the DNA molecule split apart. Each of the separated strands acts as a master mold for a new copy of its other half to form two new double strands of DNA. The new DNA strands are made from nucleotide bases that have been manufactured in the cell. By the time a cell starts to divide, all 46 chromosomes inside it have duplicated.

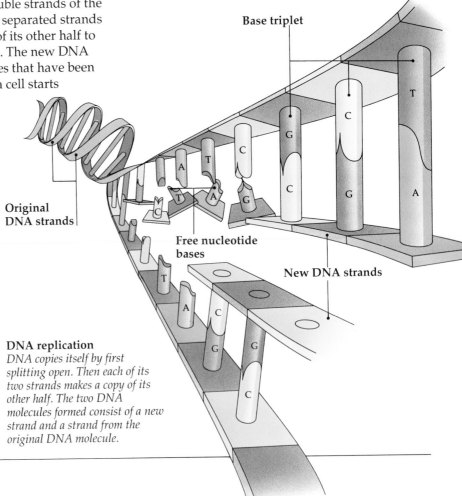

Base triplet

Original DNA strands

Free nucleotide bases

New DNA strands

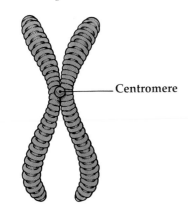

Centromere

Duplicated chromosome
After DNA replication, the amount of DNA in a cell nucleus has doubled. Just before cell division, the duplicated chromosomes are joined at a point called the centromere, which gives them an X-shaped appearance.

DNA replication
DNA copies itself by first splitting open. Then each of its two strands makes a copy of its other half. The two DNA molecules formed consist of a new strand and a strand from the original DNA molecule.

MITOSIS: CELL MULTIPLICATION BY DIVISION

Your body is constantly replacing dead or damaged cells – such as skin cells – with healthy cells, which reproduce themselves by a process of cell division called mitosis. In the process, a cell divides to form two new cells that each contain identical copies of the original cell's genetic material.

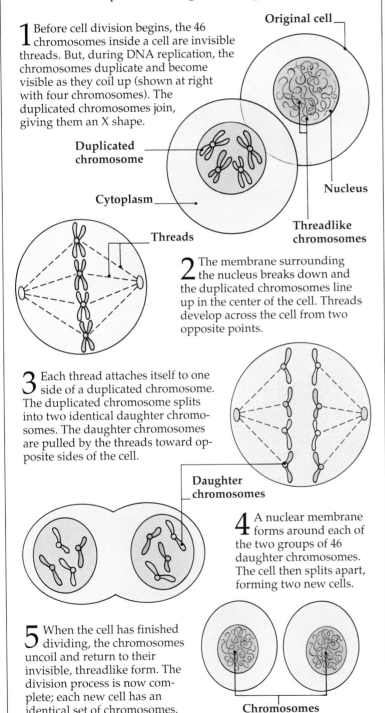

1 Before cell division begins, the 46 chromosomes inside a cell are invisible threads. But, during DNA replication, the chromosomes duplicate and become visible as they coil up (shown at right with four chromosomes). The duplicated chromosomes join, giving them an X shape.

Original cell

Duplicated chromosome

Nucleus

Cytoplasm

Threadlike chromosomes

Threads

2 The membrane surrounding the nucleus breaks down and the duplicated chromosomes line up in the center of the cell. Threads develop across the cell from two opposite points.

3 Each thread attaches itself to one side of a duplicated chromosome. The duplicated chromosome splits into two identical daughter chromosomes. The daughter chromosomes are pulled by the threads toward opposite sides of the cell.

Daughter chromosomes

4 A nuclear membrane forms around each of the two groups of 46 daughter chromosomes. The cell then splits apart, forming two new cells.

5 When the cell has finished dividing, the chromosomes uncoil and return to their invisible, threadlike form. The division process is now complete; each new cell has an identical set of chromosomes.

Chromosomes

METHODS OF REPRODUCTION

In some life-forms, such as bacteria, cell division by mitosis is used not just for growth and repair, but also to make new organisms. Individual cells break away from the parent organism and divide to form genetically identical offspring. Most higher forms of life, including humans, produce offspring by the much more complex process of sexual reproduction. In sexual reproduction, mother and father each contribute half of their genes to every child they produce. The genetic material from the two parents is packaged in special reproductive cells, which in humans are the egg and the sperm. When a sperm unites with an egg, the genes from the two parents combine into a single cell – the fertilized egg. The cell then begins a long series of divisions that will eventually form a human being with a unique genetic makeup.

Meiosis

Meiosis is the process of cell division by which the reproductive cells are formed. Unlike mitosis, which occurs in cells throughout the body, meiosis occurs only in the testicles to produce sperm and in the ovaries to produce eggs. Each egg or sperm contains only half of the genetic material of other cells because, during egg or sperm formation, the genetic material of the parent cell is divided in two. Instead of the usual 46 chromosomes, each reproductive cell has only 23.

In contrast to the identical cells formed during mitosis, each egg and sperm formed during meiosis is genetically unique. Not only does each egg or sperm cell contain only half of a person's genes, but each also contains a different selection of those genes. This genetic mixing occurs at the beginning of cell division when sections of DNA from matching pairs of chromosomes are randomly swapped (see HOMOLOGOUS CHROMOSOMES on page 39).

HOW ARE SEX CELLS FORMED?

Egg and sperm cells are formed by a type of cell division called meiosis. In meiosis, genetic material that originated from your parents is exchanged between pairs of chromosomes to produce sperm or eggs, each of which has a unique mix of genetic information. Two stages of cell division occur in meiosis to produce four new cells. In each newly formed sex cell the number of chromosomes has been reduced from 46 to 23, so that, when the sex cell combines with its counterpart from the other parent during fertilization, the resulting embryo receives half of its genes from each parent.

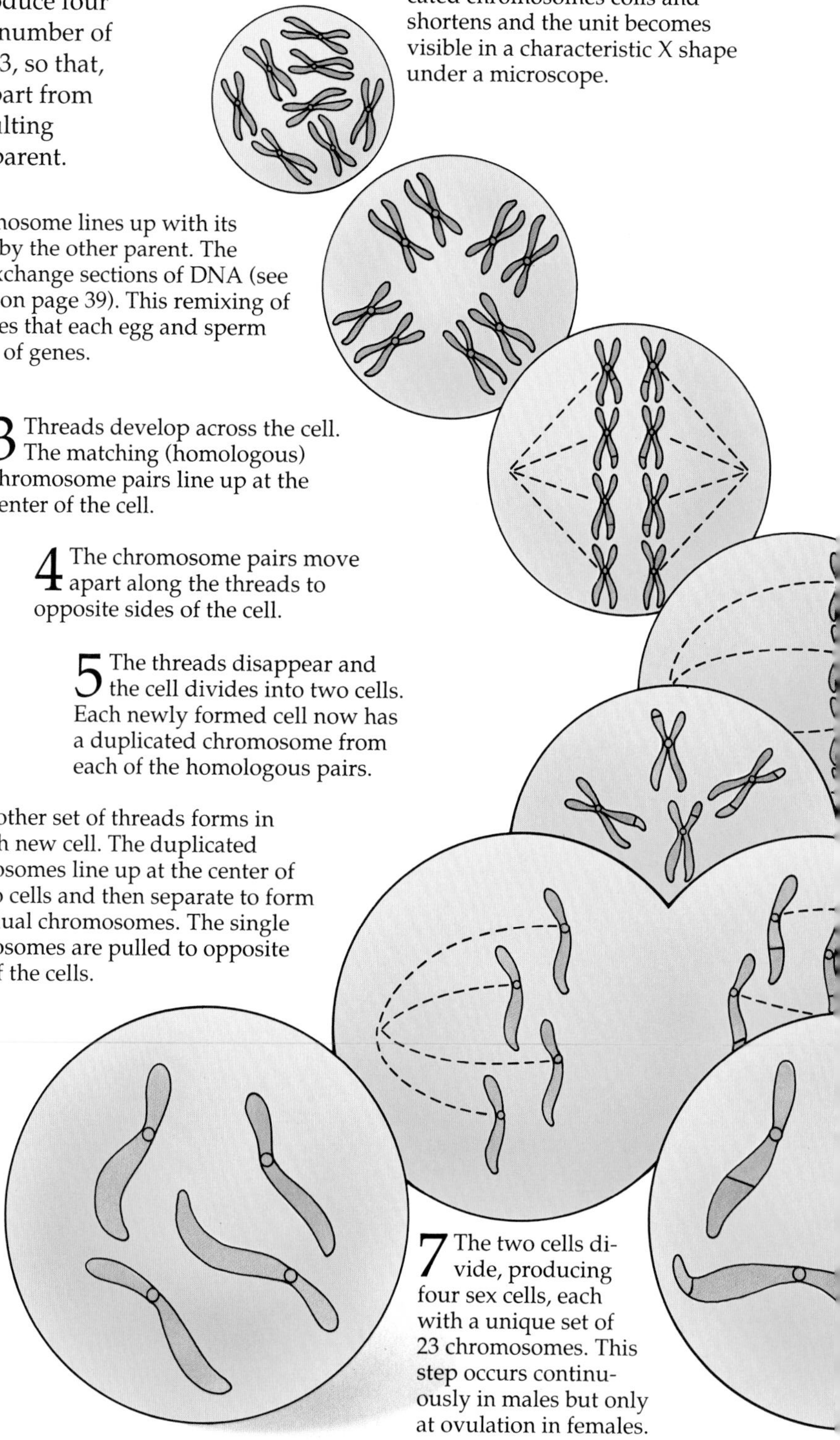

1 Before meiosis begins in a testicle or ovary, the DNA in all 46 chromosomes duplicates (only eight chromosomes are shown here). Each pair of duplicated chromosomes coils and shortens and the unit becomes visible in a characteristic X shape under a microscope.

2 Each duplicated chromosome lines up with its counterpart, supplied by the other parent. The chromosome pairs then exchange sections of DNA (see HOMOLOGOUS CHROMOSOMES on page 39). This remixing of genetic information ensures that each egg and sperm has a unique combination of genes.

3 Threads develop across the cell. The matching (homologous) chromosome pairs line up at the center of the cell.

4 The chromosome pairs move apart along the threads to opposite sides of the cell.

5 The threads disappear and the cell divides into two cells. Each newly formed cell now has a duplicated chromosome from each of the homologous pairs.

6 Another set of threads forms in each new cell. The duplicated chromosomes line up at the center of the two cells and then separate to form individual chromosomes. The single chromosomes are pulled to opposite sides of the cells.

7 The two cells divide, producing four sex cells, each with a unique set of 23 chromosomes. This step occurs continuously in males but only at ovulation in females.

Formation of sex cells

During meiosis, four sex cells are produced from one original cell. The sex cells, which contain only 23 chromosomes, mature to form egg cells in a woman and sperm cells in a man. Each sperm or egg carries a different mix of genetic material.

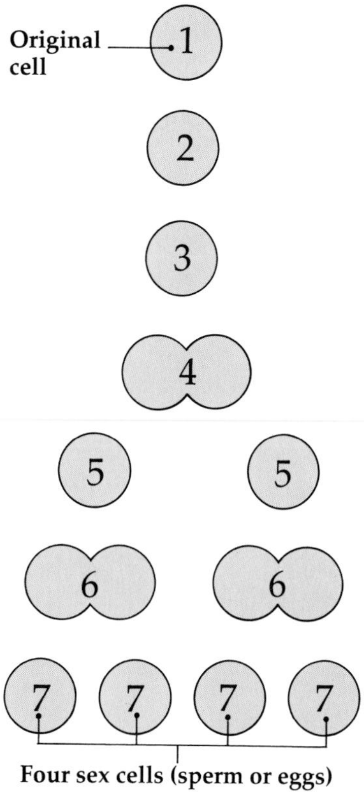

Original cell — 1

2

3

4

5   5

6   6

7 7 7 7

Four sex cells (sperm or eggs)

HOMOLOGOUS CHROMOSOMES

The 46 chromosomes in each of your cells come in 23 pairs, one member of which is provided by each of your parents. Except for the X and Y sex chromosome pair in males, the chromosomes in a pair are matching and are called homologous, from the Greek word for "agreeing." Although homologous chromosomes are similar in most respects, they are not identical. Genes at the same location on a pair of homologous chromo-somes may differ slightly in the genetic instructions that they provide for the body's construction, func-tioning, or repair. During meiosis, an exchange of genetic material between homologous chromosomes – which can occur at any point along them – results in new chromosomes with new combinations of genes, in arrangements that never occurred before and will never occur again.

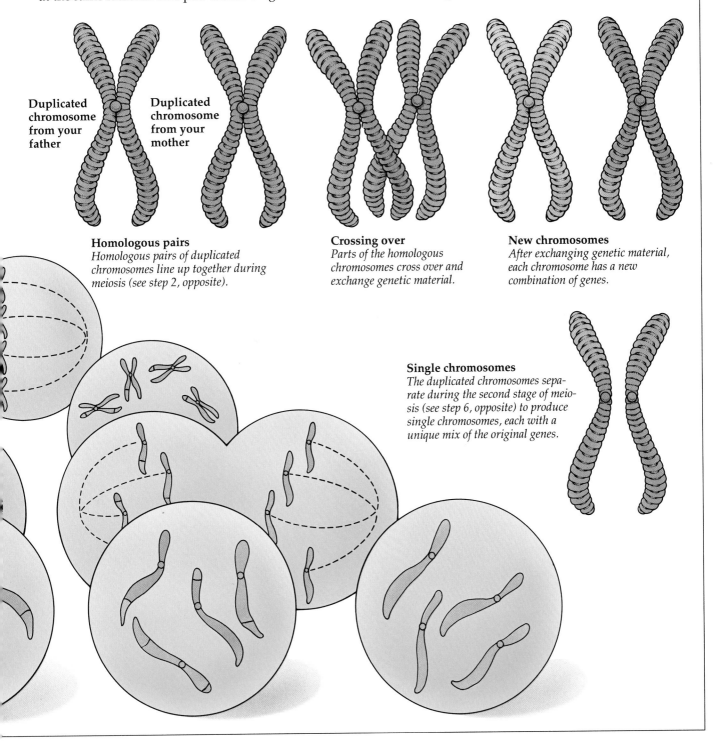

Duplicated chromosome from your father

Duplicated chromosome from your mother

Homologous pairs
Homologous pairs of duplicated chromosomes line up together during meiosis (see step 2, opposite).

Crossing over
Parts of the homologous chromosomes cross over and exchange genetic material.

New chromosomes
After exchanging genetic material, each chromosome has a new combination of genes.

Single chromosomes
The duplicated chromosomes sepa-rate during the second stage of meio-sis (see step 6, opposite) to produce single chromosomes, each with a unique mix of the original genes.

WHAT CAUSES TWINS?

Sperm and egg combine during fertilization to produce the first cell of a new human being. This new cell, the zygote, has a unique set of genes. Genetic variation is what makes you look different from your neighbors and even your closest relatives. However, because you receive half of your genes from each parent, you are likely to resemble them, and your brothers and sisters as well.

The closest genetic similarity of any two individuals is that of identical twins. Identical twins are produced when one fertilized egg divides into two separate cells. Each of the new cells divides independently, eventually producing a person. Because they develop from the same fertilized egg, identical twins have identical sets of genes. Nonidentical, or fraternal, twins develop from two separate eggs that are each fertilized by different sperm. As a result, nonidentical twins have different genetic makeups, and they resemble each other no more than any two siblings.

SEX DETERMINATION

Other than the egg and the sperm, all cells contain 23 pairs of chromosomes. Of these, the single pair of sex chromosomes determines a person's sex. Women have two X sex chromosomes in each of their cells. Men have one X chromosome and a much smaller chromosome called the Y chromosome. The sex chromosome set of a woman is described as XX; that of a man is described as XY. Female characteristics, including the ovaries, will develop in a fetus unless the Y chromosome is present. The Y chromosome stimulates the production of the male hormone testosterone, which, in turn, stimulates the formation of male characteristics, including the testicles.

TYPES OF TWINS

Women conceive identical twins at about the same rate. But a mother of one set of nonidentical twins is four times more likely than other women to have another set. Fertility drugs increase the chances of having nonidentical twins by stimulating the release of more than one egg.

Key
⚥ Male   ♀ Female

Identical twins
Identical twins develop from the same fertilized egg and they share the same placenta inside the uterus. Identical twins are always the same sex.

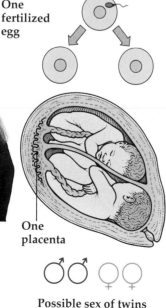

One fertilized egg

One placenta

♂♂   ♀♀
Possible sex of twins

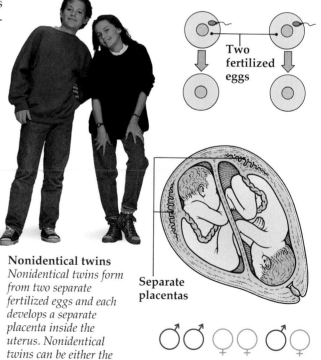

Nonidentical twins
Nonidentical twins form from two separate fertilized eggs and each develops a separate placenta inside the uterus. Nonidentical twins can be either the same or the opposite sex.

Two fertilized eggs

Separate placentas

♂♂ ♀♀ ♂♀
Possible sex of twins

Most of the cells in your body have 46 chromosomes, including two sex chromosomes. Egg and sperm cells have exactly half that number. Each egg that a woman produces has one X sex chromosome plus 22 nonsex chromosomes, or a total of 23. Each sperm that a man produces has either an X or a Y sex chromosome plus 22 nonsex chromosomes. Half of a man's sperm will carry the X chromosome; the other half will carry the Y. Because an egg contains only the X chromosome, a fetus's sex is determined by whether the sperm that fertilizes the egg carries an X or a Y sex chromosome.

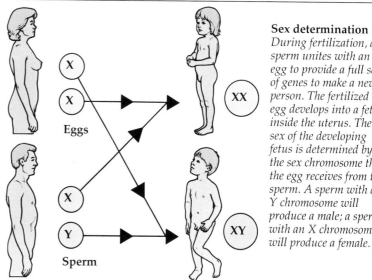

Sex determination
During fertilization, a sperm unites with an egg to provide a full set of genes to make a new person. The fertilized egg develops into a fetus inside the uterus. The sex of the developing fetus is determined by the sex chromosome that the egg receives from the sperm. A sperm with a Y chromosome will produce a male; a sperm with an X chromosome will produce a female.

THE CYCLE OF LIFE

In sexually reproducing life-forms, genes are continuously copied and passed on from generation to generation. You might say this process represents genetic immortality.

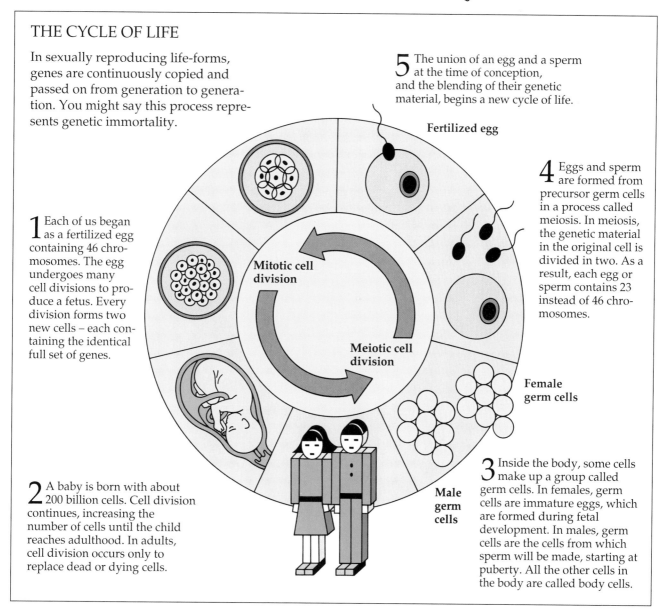

5 The union of an egg and a sperm at the time of conception, and the blending of their genetic material, begins a new cycle of life.

4 Eggs and sperm are formed from precursor germ cells in a process called meiosis. In meiosis, the genetic material in the original cell is divided in two. As a result, each egg or sperm contains 23 instead of 46 chromosomes.

1 Each of us began as a fertilized egg containing 46 chromosomes. The egg undergoes many cell divisions to produce a fetus. Every division forms two new cells – each containing the identical full set of genes.

2 A baby is born with about 200 billion cells. Cell division continues, increasing the number of cells until the child reaches adulthood. In adults, cell division occurs only to replace dead or dying cells.

3 Inside the body, some cells make up a group called germ cells. In females, germ cells are immature eggs, which are formed during fetal development. In males, germ cells are the cells from which sperm will be made, starting at puberty. All the other cells in the body are called body cells.

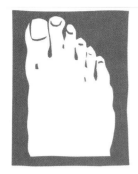

GENETIC MUTATION AND VARIABILITY

A GENETIC MUTATION is a change in the coded DNA instructions that govern a cell's activities. Any alteration or disruption in the DNA code can have a dramatic impact on a cell's functioning and that mutation will be duplicated in all the cell's descendants. Genetic mutation is the ultimate source of all biological variations. Those variations can be either beneficial or harmful.

In a gene or chromosome, a mutation is a change in the sequence of nucleotide bases that are the letters of the DNA code. Mutations result from the rare errors that occur when DNA is replicated in a cell before the cell divides. A strict sequence of bases (the code) in the DNA provides instructions for making specific proteins, which regulate the activities of cells. Any alteration in the DNA code can have a profound effect on the way in which the instructions contained in that piece of DNA are read, understood, and carried out by cells. Gene mutations can cause both helpful and harmful changes that can affect human health.

IMPACT OF MUTATIONS

The impact of a gene mutation depends on whether the affected cell is an egg or sperm (or any cell that has the potential to divide to form an egg or sperm) or one of the other cells in the body.

Mutations in body cells
A mutation in a body cell will not be passed on to children, but it will be transmitted during cell division to all descendants of the mutant cell in a person's body. Such a mutation can lead to an accumulation of abnormal cells in one area of the body. Although abnormal cells usually die or are destroyed by the immune system, they sometimes develop into a cancer.

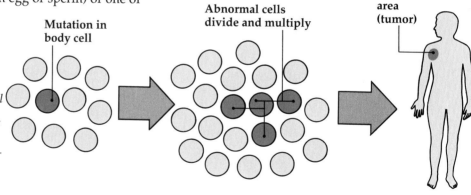

Mutation in body cell

Abnormal cells divide and multiply

Affected area (tumor)

Mutations in sex cells
A mutation in the DNA of an egg or a sperm may be transmitted to a child via that egg or sperm. The child will then carry the mutation in all of his or her cells and can, in turn, transmit the mutation to offspring. Some gene mutations have no effect or may even be beneficial. But some gene mutations can cause severe abnormalities in the children who have them.

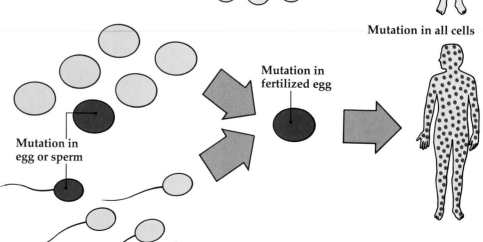

Mutation in all cells

Mutation in egg or sperm

Mutation in fertilized egg

MUTATION AND HUMAN VARIATION

One of the main principles of Darwin's theory of evolution (see page 11) is that members of all animal and plant species vary in characteristics such as size and shape. When Darwin first developed his theory, he had no clear understanding of how this natural variation occurred. But, after the discovery of chromosomes and the structure of DNA, scientists realized that the mutation of genes is the underlying source of natural variation.

Mutations and evolution

All the genetic differences between people, including physical characteristics such as eye color, are ultimately the result of mutations that have occurred in the past and have been passed down, over hundreds of thousands of years, from one generation to the next.

Most gene mutations that have been retained are beneficial, or at least are not harmful, to a person's ability to survive and reproduce. Mutations that reduce a person's chances of surviving and reproducing will die out over generations.

A persistent mutation
A good example of how a genetic mutation and its effects can persist for many generations is the "Hapsburg jaw." A large and jutting jaw and lower lip affected several members of the Hapsburg royal family of Austria and Spain between the 15th and 18th centuries, particularly the Emperor Charles V (1500-1588; above) and his great-great-grandson King Carlos II of Spain (1661-1700; above right).

Conversely, mutations that are neutral in their effects are likely to persist, and those that increase a person's chances of surviving and reproducing tend to become more common. These genetic changes are the basis of evolution.

Blood type (see page 52) is one example of human variation that has probably resulted from harmless gene mutations. The relative frequency of the different genes that determine blood type varies noticeably from one part of the world to another. Scientists think that some genes for blood type may have originated in geographically restricted areas of the world as the result of mutation, and were subsequently distributed to new populations by migration.

Other sources of variation

Although mutation is the primary source of genetic variation among people, many other processes contribute as well.

Because sexual reproduction involves the mixing and combining of genes from different people, it is a powerful mechanism for increasing genetic diversity. Through sexual reproduction, genes are continuously being remixed to produce new combinations of genes.

Human differences and similarities
Genetic variation is responsible for many noticeable differences among people, such as the color of their hair, skin, and eyes. Studies of a wider range of human characteristics, including body chemistry, have revealed that, for the most part, people are genetically very similar.

HOW DO GENE MUTATIONS OCCUR?

A gene mutation occurs as the result of a chemical change in a gene. In genes, each group of three nucleotide bases – called a base triplet – represents the code for one amino acid. The order in which amino acids are strung together determines which protein they will form. One gene, therefore, provides the instructions for the production of one protein. An alteration in the sequence of bases in the DNA code can result in the production of a defective protein, which can interfere with body functioning. The effects of these changes in the DNA code vary widely, from minor oddities such as hairy elbows to serious genetic disorders such as cystic fibrosis.

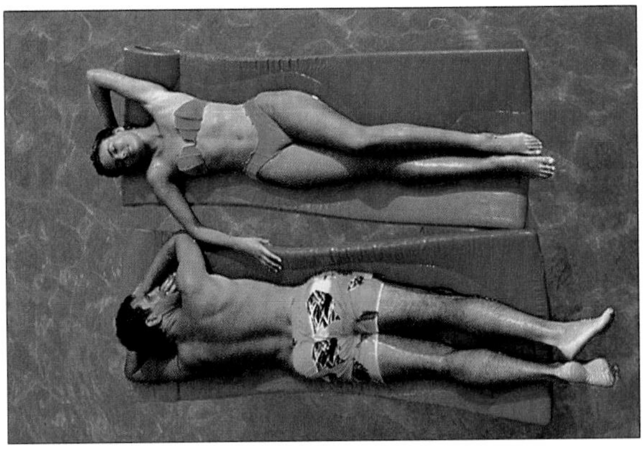

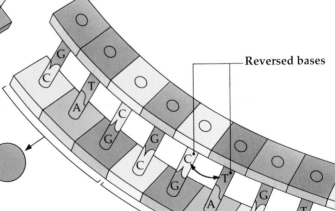

SOURCES OF GENE MUTATIONS

Most gene mutations happen spontaneously. However, exposure to certain environmental poisons can speed up the rate at which some mutations occur. Chemicals, such as mustard gas (which was used during World War I), and some forms of radiation, such as ultraviolet light, are known to cause gene mutations.

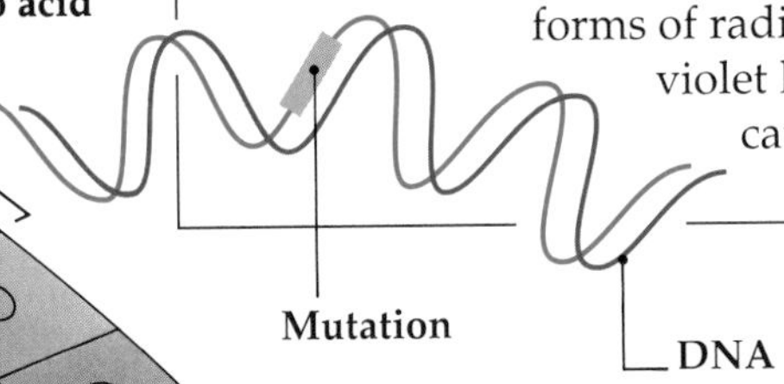

Mutation

DNA

Normal gene function
In a normal gene, each base triplet is the code for a specific amino acid, which joins together with other amino acids to make a protein.

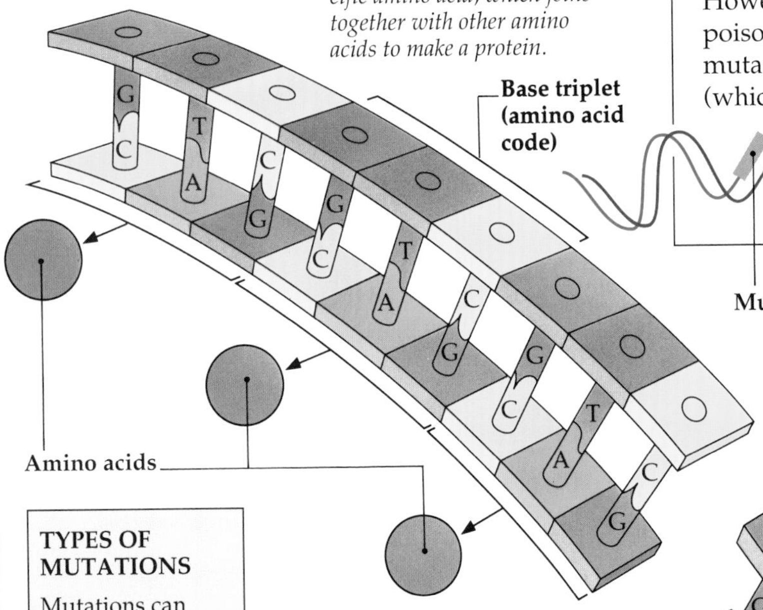

Base triplet (amino acid code)

Amino acids

TYPES OF MUTATIONS

Mutations can occur in single genes (shown here) or on chromosomes (see STRUCTURAL ABNORMALITIES on page 64). Gene mutations can result when just a tiny part (usually one base pair) of the DNA code is altered. These so-called point mutations account for many genetic disorders, including hemophilia.

Inversion
An inversion is a gene mutation in which two or more base pairs are reversed. Depending on its position along the DNA strand, this change in the order of the bases may cause a change in one or two amino acids.

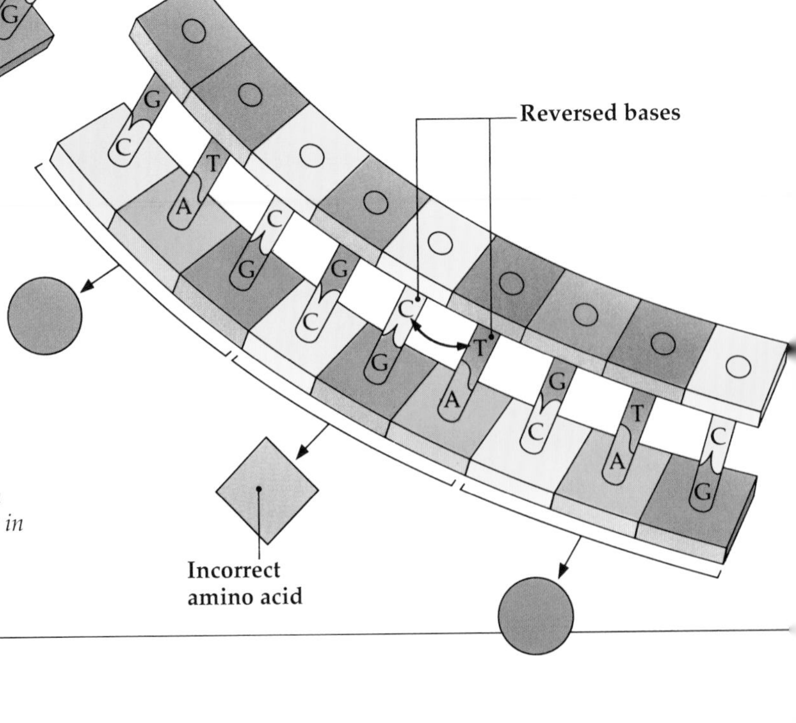

Reversed bases

Incorrect amino acid

DNA PROOFREADERS

An error in the order of bases in a gene does not necessarily cause a problem. Cells have a variety of special proofreading enzymes that can spot and correct errors in the DNA code. But if the gene controlling a proofreading enzyme contains a mutation, the gene will produce an enzyme that cannot do its job. Although the DNA proofreaders only miss about one in 100,000 mistakes, 10 times as many errors might go unnoticed if the enzymes are not working. One such abnormal repair enzyme makes a person with a particular gene mutation susceptible to skin cancer after prolonged exposure to the sun.

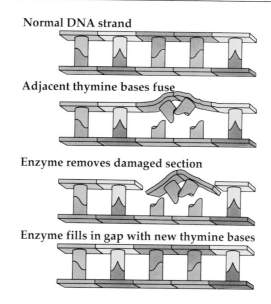

Normal DNA strand

Adjacent thymine bases fuse

Enzyme removes damaged section

Enzyme fills in gap with new thymine bases

Preventing mutations

Ultraviolet light can cause two adjacent thymine (T) bases in a length of DNA to fuse together. This fusion leads to a mutation unless a proofreading enzyme contained in the cell detects the error and corrects it. This enzyme can remove the fused thymine bases from the DNA and replace them with two new ones, thereby restoring the DNA to its original form.

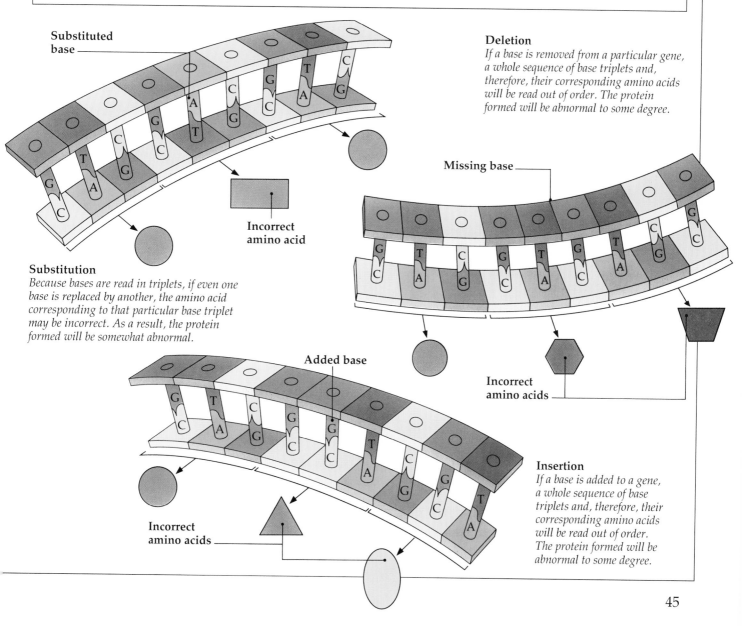

Substituted base

Substitution

Because bases are read in triplets, if even one base is replaced by another, the amino acid corresponding to that particular base triplet may be incorrect. As a result, the protein formed will be somewhat abnormal.

Incorrect amino acid

Deletion

If a base is removed from a particular gene, a whole sequence of base triplets and, therefore, their corresponding amino acids will be read out of order. The protein formed will be abnormal to some degree.

Missing base

Incorrect amino acids

Added base

Incorrect amino acids

Insertion

If a base is added to a gene, a whole sequence of base triplets and, therefore, their corresponding amino acids will be read out of order. The protein formed will be abnormal to some degree.

Historically, the migration of populations to new geographical areas probably led to the emergence of genes that helped people adapt to their environment. Variations in characteristics such as skin color and body build among people living in different parts of the world today may represent such environmental adaptations. For example, the dark skin pigmentation of people living in tropical countries may protect them from the damaging effects of the sun. But sunlight also has the beneficial effect of stimulating skin cells to produce vitamin D, which provides calcium and builds bones. The light skin of people living at higher latitudes with less intense sunlight appears to be an adaptation that helps their skin cells make a sufficient amount of vitamin D despite the limited sunlight.

Random fluctuations occur in the frequency of some genes in a population from one generation to the next. Gene frequency is the incidence of a certain form of a gene (such as the brown-eye form of the eye-color gene) compared with the total of all forms (such as all possible eye colors) of that gene. Random changes in gene frequency are most noticeable in small, isolated populations. Over generations, increases in the frequency of some genes tend to make people in these populations more similar than they would otherwise be.

How much do people vary genetically?

In recent years, new techniques for analyzing genes have made it possible to evaluate more accurately the extent of genetic diversity among people. Researchers have found that the genetic differences among populations are much smaller than expected. In fact, three out of four of the genes of any two people anywhere in the world are similar. It is only the other 25 percent of genes (such as the genes for the color of hair and eyes and for blood type) that vary noticeably from person to person. Siblings, because they have the same parents, are even more similar. They share not only the genes that most people have in common, but they also have identical copies of the genes that usually differ among people. As a result, siblings can be expected to have considerably more than 75 percent of the same genes. The genes of identical twins, who originate from the same fertilized egg, are all identical.

GENETIC ISOLATION

Members of some isolated communities – such as certain religious sects and the inhabitants of remote islands – marry only among themselves. After several generations, such groups tend to lose their genetic diversity because the variety of genes within the group gradually diminishes. The smaller the population, the greater the chance that certain genes will randomly increase or decrease in number over many generations. If one or more founding members of a community are carriers of a particular gene, that gene may, over generations, become more common in the community than it would be in the general population. This is more likely to occur when one generation of a family with the gene has a higher-than-average number of children who then also have children.

Affected communities
Isolated groups of people, such as the Amish community in Pennsylvania, have been affected by disorders caused by genes that originated in the founding members of the group. Because the Amish marry only inside the group, these disease genes tend to increase in the population.

SICKLE CELL TRAIT

Sickle cell anemia, which occurs mainly in blacks, is an inherited blood disease caused by an abnormality in the structure of hemoglobin, the oxygen-carrying pigment in red blood cells. The disorder occurs in people who inherit the gene mutation from both of their parents. Sickle cell anemia is no longer always fatal in childhood, but it still causes serious medical problems. If the disease gene is inherited from only one parent, most of the hemoglobin in the red blood cells is normal and the person has a condition called sickle cell trait, which causes few health problems. The sickle cell trait appears to provide some protection against falciparum malaria; scientists believe this is why the mutant sickle cell gene is common in parts of Africa where malaria is present.

Healthy red blood cells   **Sickle-shaped red blood cells**

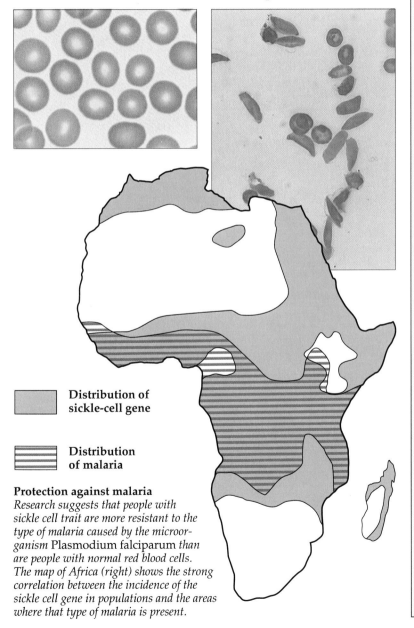

Distribution of
sickle-cell gene

Distribution
of malaria

Protection against malaria
Research suggests that people with sickle cell trait are more resistant to the type of malaria caused by the microorganism Plasmodium falciparum *than are people with normal red blood cells. The map of Africa (right) shows the strong correlation between the incidence of the sickle cell gene in populations and the areas where that type of malaria is present.*

ASK YOUR DOCTOR
MUTATIONS

Q **My son has been diagnosed as having hemophilia. I thought hemophilia was hereditary, but nobody in our family has ever had it. How could this happen?**

A Hemophilia is caused by a gene carried on the X sex chromosome. It almost always occurs in males, who inherit the harmful gene on their mother's X chromosome. Most affected families have a history of the disease. But about one third of the time, hemophilia results from a new mutation that occurred during the formation of a parent's egg or sperm cell. This new mutation may be transmitted, in the next generation or a subsequent generation, to a boy who will have the disorder.

Q **Can new gene mutations be detected during pregnancy?**

A Usually not. Because new gene mutations cannot be predicted, it is not possible to identify those pregnancies that may be at risk. In addition, a large number of different genes can mutate and cause genetic disorders; this makes it impossible to test during pregnancy for every single disorder that might result from a new gene mutation.

Q **How often do gene mutations occur in our cells?**

A Mutations that have any noticeable or harmful effect are relatively rare. The mutation rates of different genes vary, but the estimated risk that an individual gene will mutate during the formation of an egg or sperm cell ranges from one in 100,000 to one in a million. Considering the complexity of the DNA copying process, this is a remarkably low error rate.

PATTERNS OF HEREDITY

HEREDITY IS THE TRANSMISSION of characteristics from parents to children through the influence of genes. The basic laws of heredity were first described in the 1860s, long before the chemical nature of genes was known. Our understanding of the complexity of genes and their role in heredity advanced rapidly after the discovery in 1953 of the structure of DNA (deoxyribonucleic acid), the fundamental carrier of genetic information.

The genes in a cell are packaged in long strands of DNA called chromosomes. Inside a chromosome, genes are arranged in a precise order. The genes controlling most characteristics come in pairs – one inherited from the father and one from the mother. The paired genes are located at corresponding sites on 22 pairs of chromosomes called autosomes. The sex chromosomes X and Y make up the remaining chromosome pair. Females have two X sex chromosomes; males have one X and one Y. Every one of your cells contains a complete set of all of your genes. Each set is packaged in 23 pairs of chromosomes – 46 chromosomes in all.

The heredity of some traits, such as eye color, is determined by a single pair of genes. Other, more complex traits, such as intelligence, result from combinations of various genetic and environmental influences.

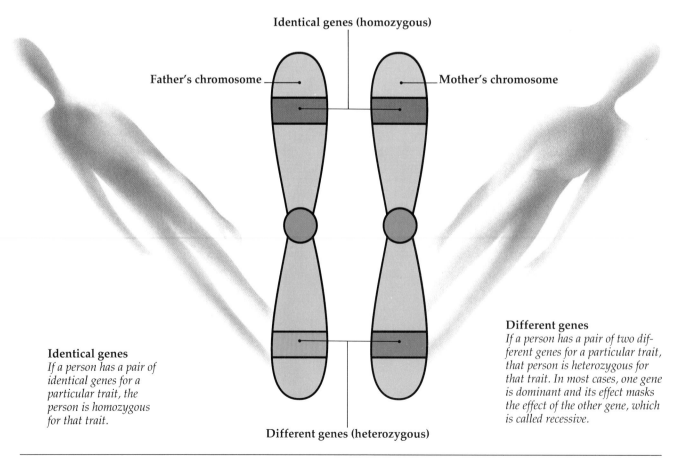

Identical genes (homozygous)

Father's chromosome

Mother's chromosome

Identical genes
If a person has a pair of identical genes for a particular trait, the person is homozygous for that trait.

Different genes
If a person has a pair of two different genes for a particular trait, that person is heterozygous for that trait. In most cases, one gene is dominant and its effect masks the effect of the other gene, which is called recessive.

Different genes (heterozygous)

DOMINANT AND RECESSIVE CHARACTERISTICS

Many traits are determined by a single pair of genes. The members of the pair may be either dominant or recessive. Because a dominant gene overrides a recessive gene, only one dominant gene is required to produce its effect on a person. But a person must have two copies of a recessive gene to exhibit the characteristic determined by that gene. For example, in simplified terms, eye color is regulated by a single pair of genes, either of which may be a gene for blue eyes or a gene for brown eyes. The gene for brown eyes is dominant and the gene for blue eyes is recessive. Therefore, a person with brown eyes may have either one or two copies of the dominant gene for brown eyes. A person has blue eyes only if he or she has inherited two copies of the recessive gene for blue eyes.

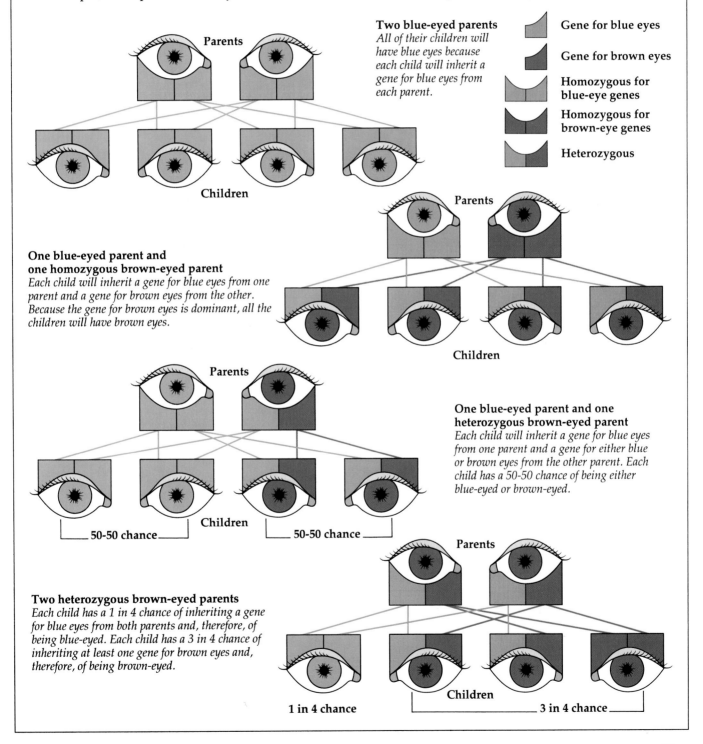

Two blue-eyed parents
All of their children will have blue eyes because each child will inherit a gene for blue eyes from each parent.

Gene for blue eyes

Gene for brown eyes

Homozygous for blue-eye genes

Homozygous for brown-eye genes

Heterozygous

Parents

Children

One blue-eyed parent and one homozygous brown-eyed parent
Each child will inherit a gene for blue eyes from one parent and a gene for brown eyes from the other. Because the gene for brown eyes is dominant, all the children will have brown eyes.

Parents

Children

Parents

Children

One blue-eyed parent and one heterozygous brown-eyed parent
Each child will inherit a gene for blue eyes from one parent and a gene for either blue or brown eyes from the other parent. Each child has a 50-50 chance of being either blue-eyed or brown-eyed.

50-50 chance

50-50 chance

Two heterozygous brown-eyed parents
Each child has a 1 in 4 chance of inheriting a gene for blue eyes from both parents and, therefore, of being blue-eyed. Each child has a 3 in 4 chance of inheriting at least one gene for brown eyes and, therefore, of being brown-eyed.

Parents

Children

1 in 4 chance

3 in 4 chance

SIMPLE INHERITANCE

Some characteristics, such as blood type, the color of eyes and hair, and even the ability to taste certain substances, are controlled by a single pair of genes located on autosomes (nonsex chromosomes). The transmission of these traits follows the basic laws of heredity first described by Gregor Mendel in his experiments with pea plants in the 1860s (see page 12). The inheritance of hair color, for example, is usually determined by a single pair of genes. The hair-color genes can be in either of two principal forms, one for dark hair and the other for light hair. When a person has a copy of each of the two forms, the instructions provided by the gene for dark hair (the dominant gene) override the instructions provided by the gene for light hair (the recessive gene). To be dark-haired, a person needs to inherit only one gene for dark hair from one parent, while, to be light-haired, he or she must inherit two copies of the gene for light hair, one from each parent. As a result, dark hair is considered the dominant trait and light hair is the recessive trait. Similar rules apply to many other characteristics that are controlled by a single pair of genes, including those shown on this page and on page 51.

Curly hair
Like dark hair, curly hair is a dominant trait. A person with curly hair may have only one dominant gene for curly hair and a recessive gene for straight hair, or two copies of the dominant curly hair gene.

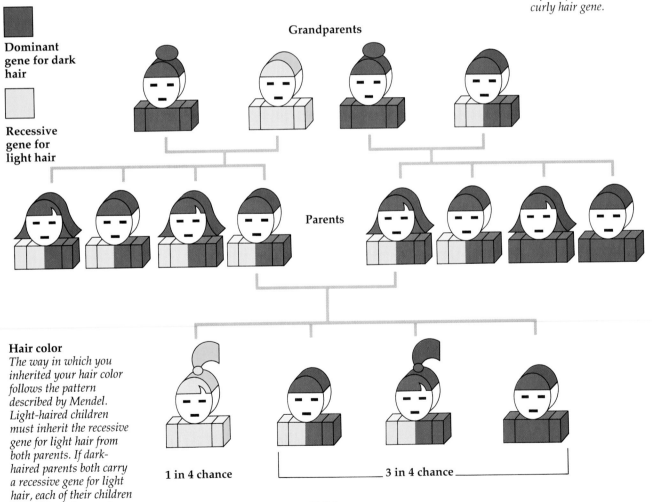

Dominant gene for dark hair

Recessive gene for light hair

Grandparents

Parents

Hair color
The way in which you inherited your hair color follows the pattern described by Mendel. Light-haired children must inherit the recessive gene for light hair from both parents. If dark-haired parents both carry a recessive gene for light hair, each of their children has a 1 in 4 chance of having light hair.

1 in 4 chance

3 in 4 chance

Children

Albinism

Albinism, a condition that results from the body's inability to make sufficient amounts of a pigment called melanin, is a good example of a recessive disorder. Albinos have little or no pigment in their skin and hair, making them extremely sensitive to the damaging effects of the sun. An affected person has inherited two copies of the recessive gene for albinism, one from each parent.

Ability to roll your tongue

The ability to roll your tongue (left) is determined by a dominant gene that controls certain muscles in your tongue. If you have either a single or double copy of the gene, as do about 85 percent of people, you can roll your tongue easily. If you do not have at least one copy of the gene, you will be unable to roll your tongue.

Taste

Geneticists think that the ability to taste some substances may be inherited. In an accidental discovery, scientists found that a chemical compound called phenylthiocarbamide can be tasted only by people who have one or two copies of a particular dominant gene that makes the chemical taste bitter. More than 70 percent of Americans can taste the chemical.

FAMILY INHERITANCE PATTERNS

The pattern of inheritance of a trait in a family from generation to generation can help doctors to identify hereditary disorders and the way in which they are transmitted from parents to child. The example below shows the inheritance pattern in one family of a dominant disorder called polydactyly. A person with polydactyly is born with an extra finger or toe.

☐ **Normal male**
■ **Affected male**
○ **Normal female**
● **Affected female**

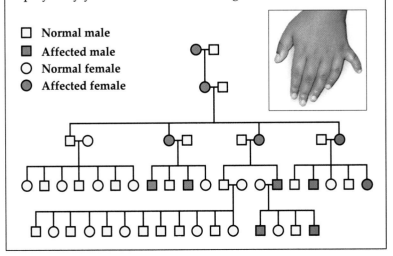

ASK YOUR DOCTOR
HEREDITY

Q **Both my wife and I have dark hair; our daughter has light hair. Since there is a 1 in 4 chance of our having a light-haired child, does this mean that any other children we have will be dark-haired?**

A No. The statistic applies to every child that you have. Each of your children has a 1 in 4 chance of being light-haired, regardless of the hair color of any previous children. Your next child may have either light hair or dark hair, but there is a significantly greater chance (75 percent) that he or she will have dark hair.

Q **My husband and I both have brown eyes and, so far, we have two brown-eyed children. Does this mean we will never have any blue-eyed children?**

A It is impossible to say for sure without knowing the eye color of your parents and your wife's parents. In addition to the dominant brown-eye gene, you and your wife may each carry the recessive gene for blue eyes. If that is the case, each of your children has a 1 in 4 chance of having blue eyes.

Q **Our child is an albino but my wife and I are not. If this condition is inherited, how could this have happened?**

A Because albinism is a recessive disorder, a child must inherit a copy of the recessive gene from both parents to be noticeably affected by it. This means that both you and your wife carry the gene for albinism. In your cases, the recessive gene has been neutralized by the presence of a normal, dominant gene. But each child that you have has a 1 in 4 chance of being an albino.

HOW IS BLOOD TYPE INHERITED?

Using the ABO blood-grouping system, blood type is classified by the presence or absence of different types of molecules, called antigens, on the surface of red blood cells. People whose cells carry type A antigens have type A blood; people with type B antigens have type B blood. If their cells carry both A and B antigens, their blood type is AB. People whose cells do not carry any antigens have type O blood. The genes in a pair that determine a person's blood type can exist in any of three forms – A, B, or O. Blood type can be further broken down into Rh positive and Rh negative, depending on the presence of other antigens.

How many possible blood types are there?

Each member of a pair of genes that determines your blood type may be either A, B, or O. A and B are equally dominant. As a result, they are exhibited together as type AB. Because the O gene is recessive, you need two copies of it to be blood type O. The gene combinations for the four blood types are shown at right.

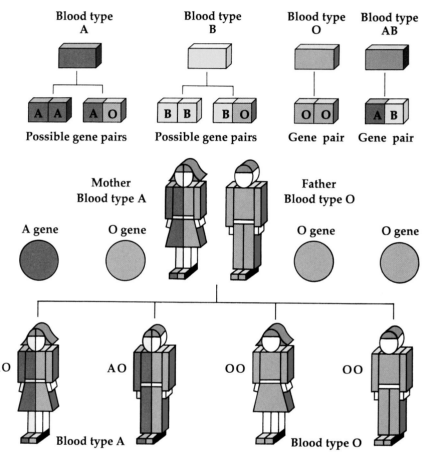

Blood type A — Possible gene pairs: A A, A O

Blood type B — Possible gene pairs: B B, B O

Blood type O — Gene pair: O O

Blood type AB — Gene pair: A B

Inheritance of blood type

You inherit the genes that determine your blood type from your parents, one gene from your father and one from your mother. For example, if your mother has the A and O genes (making her blood type A) and your father has two O genes (making him blood type O), you have a 50 percent chance of inheriting one A and one O gene, making you type A. You also have a 50 percent chance of inheriting two O genes, making you type O (see right).

Mother Blood type A — A gene, O gene

Father Blood type O — O gene, O gene

AO — Blood type A

AO — Blood type A

OO — Blood type O

OO — Blood type O

PREDICTING BLOOD TYPE

Because blood type is determined by genes, the possible blood types of children (center row) can be predicted if the blood type of both parents is known. The occurrence of some blood types among these children can also be ruled out (bottom row). Ruling out blood type has been used in paternity testing to exclude a man as the possible father of a child.

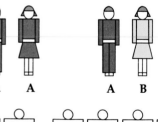
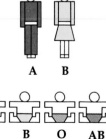

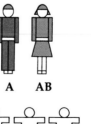
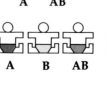
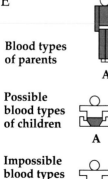

Blood types of parents

A   A | A   B | A   AB | A

Possible blood types of children

A   O | A   B   O   AB | A   B   AB | A

Impossible blood types of children

B   AB | | O | B

BLOOD TRANSFUSIONS

If you receive a transfusion of blood of a blood type different from your own, your immune system may recognize antigens (protein markers) on the surface of the donor red blood cells as foreign and launch an attack against them. This defensive response provokes a severe reaction. People with type O blood are considered universal donors because their red blood cells do not carry either A or B antigens. As a result, type O blood seldom produces an immune response in a recipient. People who have type AB blood are considered universal recipients because their red blood cells carry both the A and B antigens. Therefore, the presence of these antigens on donor blood of any type is not regarded as foreign and produces no immune response.

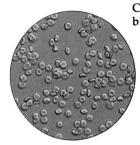

Compatible blood

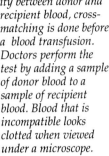

Crossmatching blood
To detect incompatibility between donor and recipient blood, crossmatching is done before a blood transfusion. Doctors perform the test by adding a sample of donor blood to a sample of recipient blood. Blood that is incompatible looks clotted when viewed under a microscope.

Incompatible blood

PATERNITY TESTING

An unintended but useful by-product of the classification of blood type is its application to cases of disputed paternity. For example, if a child has type O blood, he or she must have inherited a copy of the O blood type gene from each parent. Therefore, if the child's mother is blood type A, she must have one O and one A gene. The father must also have an O gene and, therefore, cannot be of blood type AB. Although such blood tests cannot prove definitively that a man is the father, they can rule out a person if neither he nor the child's mother has the child's blood type. The technique of DNA fingerprinting (see page 113) is now replacing blood-type analysis in paternity testing because it offers a much higher probability of linking the biological father to a child.

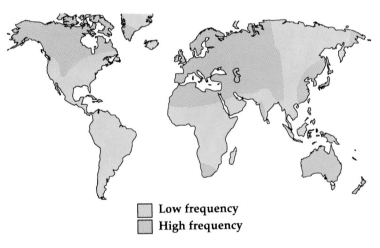

Distribution of blood type A
The frequency of blood types varies among geographically separate populations. The map at left shows the distribution of type A blood among various native populations. Scientists think the variation is caused by migration. If members of an immigrant group with different blood type patterns have children with members of the native population, the frequency of certain blood types in the native population may change.

☐ Low frequency
☐ High frequency

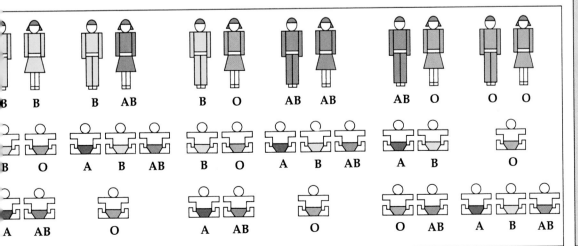

SEX-LINKED INHERITANCE

The inheritance of characteristics that are determined by genes located on the sex chromosomes is called sex-linked inheritance. Sex-linked traits usually affect one sex more than the other.

Females have two X sex chromosomes, one inherited from each parent. Males have an X sex chromosome that comes from their mother and a Y sex chromosome from their father. Almost all sex-linked traits are governed by genes located on the X chromosome and are called X-linked traits. Other than male characteristics, the only trait thought to be dictated by genes located on the Y chromosome is the unusual one of hairy ears.

X-linked recessive inheritance

Most X-linked traits and disorders are recessive. Because females have two X chromosomes, the effects of a recessive X-linked gene are usually neutralized by the effects of a corresponding dominant gene on the other X chromosome. In males, however, the recessive gene produces an effect because males have only one X chromosome – the Y chromosome does not carry a corresponding dominant copy of the gene to neutralize its effect. For this reason, X-linked recessive traits affect mostly males. A male almost always inherits an X-linked trait or disorder from his mother, who is a carrier of the recessive gene but is herself usually unaffected. Color-blindness is an X-linked trait (see right). A number of serious disorders, including hemophilia and Duchenne type muscular dystrophy, also follow an X-linked recessive pattern of inheritance (see page 80).

In unusual circumstances, an X-linked recessive trait or disorder may affect a female. For example, a girl whose father is color-blind and whose mother is an unaffected carrier of the gene for color-blindness may inherit an X chromosome with the abnormal gene from both parents – and she will be color-blind.

COLOR-BLINDNESS

A person who is color-blind has difficulty seeing the difference between reds and greens. Like most X-linked recessive traits, color-blindness is more common in men than in women. When a man who is color-blind and a woman with normal genes for color vision have children, none of their children will be color-blind, but all of their daughters will be carriers of the abnormal gene. When a woman carrier and a man with normal color vision have children, each of their sons has a 50 percent chance of being color-blind and each daughter has a 50 percent chance of being a carrier.

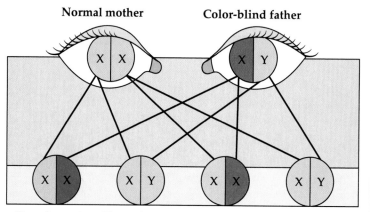

Color-blind father
If a man who is color-blind has children with a woman who has normal genes for color vision, all of their daughters will receive the affected X chromosome from their father. But, because they will also receive a normal X chromosome from their mother, they will be unaffected carriers of the gene. Because boys receive only the Y chromosome from their color-blind father and an X chromosome from their normal mother, they are not affected.

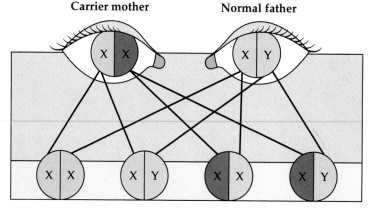

Carrier mother
If a female carrier and a man with normal vision have children, each of their daughters has a 50 percent chance of receiving the mother's affected X chromosome and of also being a carrier. Each son also has a 50 percent chance of receiving her affected chromosome but, because its effect is not neutralized by a normal X chromosome, he will be color-blind.

X-linked dominant inheritance

An X-linked dominant trait is one that is determined by a dominant gene on an X chromosome. X-linked dominant traits are rare – one of the few examples is a type of tooth discoloration.

Each child of a woman with an X-linked dominant disease gene has a 50 percent chance of inheriting her affected X chromosome and having the disease. Only the daughters of an affected man will inherit his X chromosome. This father-daughter inheritance pattern distinguishes an X-linked dominant trait from other dominant traits.

Y-linked inheritance

The way in which genes located on the Y chromosome are inherited is simple. Because only males have a Y chromosome,

fathers transmit their Y chromosome only to their sons. Therefore, a man with a trait regulated by a gene on the Y chromosome will pass the trait on to all of his sons but to none of his daughters, because they inherit only his X chromosome. Apart from providing for maleness, little is actually known of the Y chromosome's role in heredity.

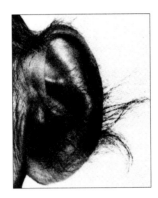

Y-linked characteristics
Other than maleness, the only trait that is thought to be Y-linked is hairy ears. Men with the Y-linked gene have long, stiff hairs in their ears, as shown at left. The trait is transmitted from father to son.

SEX-LIMITED TRAITS

Sex-limited traits affect structures or processes that exist in only one sex. Even though the genes responsible for these traits are not carried on the sex chromosomes, they almost always occur in one sex only. For example, baldness is a condition that appears to be governed by a single dominant gene located on one of the nonsex chromosomes. Because the level of activity of the gene and the degree of baldness are influenced by the presence of male sex hormones, the trait is limited to males.

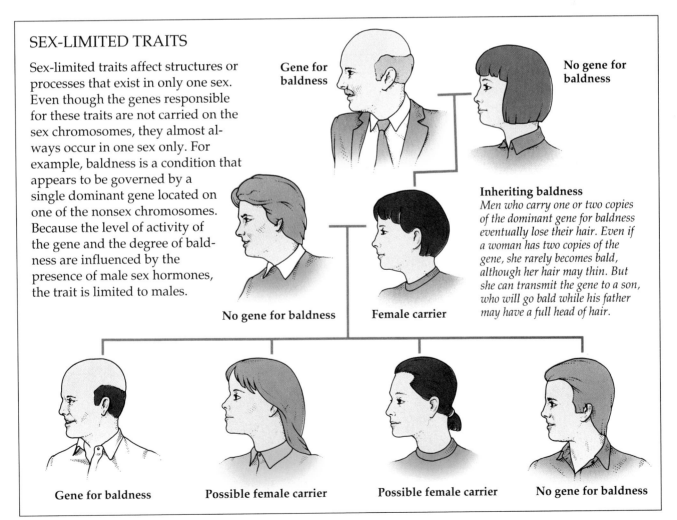

Gene for baldness

No gene for baldness

No gene for baldness

Female carrier

Inheriting baldness
Men who carry one or two copies of the dominant gene for baldness eventually lose their hair. Even if a woman has two copies of the gene, she rarely becomes bald, although her hair may thin. But she can transmit the gene to a son, who will go bald while his father may have a full head of hair.

Gene for baldness

Possible female carrier

Possible female carrier

No gene for baldness

COMPLEX INHERITANCE

The inheritance pattern of some characteristics, such as height and skin color, is difficult to determine because they are thought to result from the interaction of many different genes and environmental factors. For some of these traits, genes are the primary influence and environment plays a more minor role; for other traits, the relative importance of these influences may be reversed.

Skin color

Skin color is one of the best examples of a trait that is governed by many factors. There are at least 36 shades of skin color, ranging from almost black to almost white. Geneticists believe that variation in skin color among populations throughout the world is the result of the interaction of at least four genes. The intensity of skin pigmentation tends to be greatest in people who come from or live in areas close to the equator, while people who live farther from the equator tend to have lighter skin. This genetically determined variation in skin tone is probably the result of evolutionary adaptation to varying degrees of sunlight. But environment can also have an effect on the color and texture of your skin. For example, exposure to the sun makes your skin turn darker and, after prolonged exposure, become dry and wrinkled.

Height

Your height is also thought to be regulated by several genes. It appears that some genes ("tall" genes) strongly encourage growth, while other genes ("short" genes) encourage it less. The height you reach depends on the ratio of tall to short genes that you received from your parents. It is unlikely that a child will inherit all of one type of gene from both parents and, therefore, be exceptionally tall or exceptionally short. A child usually inherits a mixture of tall and short genes and reaches an average height. However, a child of tall parents tends to inherit more tall genes than short genes. Environmental influences can also affect a person's growth. Eating a poor diet, for example, can prevent a person from reaching his or her maximum potential height.

IMPRINTED GENES

In the mid 1980s, geneticists discovered that some genes, not just those carried on the X or Y sex chromosomes, have different effects on people, depending on whether the genes come from their mother or their father. These are known as imprinted genes. Scientists believe that there may be as many as 35 genetic disorders – including insulin-dependent diabetes, epilepsy, congenital heart disease, early-onset Huntington's chorea, and some childhood cancers – that are caused by imprinted genes. The full significance of this new discovery is still being evaluated.

Distribution of height

Height is thought to be determined by several genes. In most groups of people, individuals range from very short to very tall, with the majority of people falling in the middle. The average height of one population can differ dramatically from that of another, as shown in the graphs at right. The graphs compare the frequency of different heights for Pygmies from Zaire, Americans, and Dinkas from Sudan.

Pygmies

Americans

Frequency

Dinkas

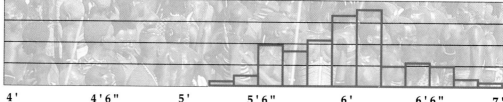

Height   4'   4'6"   5'   5'6"   6'   6'6"   7'

Whorl

Loop

Arch

Fingerprints: no two are alike

Fingerprint pattern, which is a trait influenced by several genes, has been studied extensively. The three basic fingerprint patterns – the arch, the loop, and the whorl – are shown here. No two people – not even identical twins – have identical fingerprints. Fingerprint patterns are determined by an unknown number of genes that interact with the environment inside the uterus during fetal development.

Body build

Body build is an example of a trait of complex inheritance that probably represents evolutionary adaptation to the environment. Although there are exceptions to the rule, many differences in body build among populations of the world are thought to be influenced by the climate in which they live. For example, Eskimos tend to be stout, which helps them conserve body heat (right). Conversely, some Africans are tall and thin, which provides them with more body surface to help them dissipate heat (above right).

ASK YOUR DOCTOR
HEREDITY

Q **I recently read that you can inherit a predisposition to alcoholism. Is this true?**

A Researchers are gathering an increasing amount of evidence indicating that genes play an important role in susceptibility to alcoholism. However, a person with the genes for alcoholism will obviously not become an alcoholic unless he or she drinks. Most geneticists believe that alcoholism results from a complex interaction of one or more genes and environmental factors.

Q **My parents are both extroverted, strong-willed people, and so am I. Does that mean I inherited my personality from them?**

A Like other complex characteristics, your personality is regulated by the interplay of many of your genes and the influences of your environment. Because the inheritance of personality does not follow a simple pattern, it is not easy to determine. Although genes are thought to play a part, many geneticists believe that personality is influenced much more by the way people are raised and their life experiences.

Q **My father and his father both had full heads of hair into old age, but I am becoming bald at 30. If baldness is inherited, from whom did I inherit it?**

A You probably inherited the gene for baldness from your mother. Women can carry the gene without any noticeable effect because their hormones protect them from extreme hair loss. If your father had the gene, he would most likely have been bald because the male sex hormones make the gene active.

CHAPTER THREE

GENES AND DISEASE

INTRODUCTION

CHROMOSOME
ABNORMALITIES

SINGLE-GENE
DISORDERS

GENETICS OF
COMMON DISEASES

BIRTH DEFECTS

GENES AND CANCER

MANY OF THE CHRONIC diseases that affect people today are determined, at least in part, by genes. It is difficult to know the true incidence of diseases that are influenced by genetic factors because many disorders do not appear until late in a person's life. By age 25, nearly 5 percent of the population are affected by disorders in which genes play a large part. Up to 60 percent are affected by such disorders at some time in their lives. About one in 30 babies has a birth defect, a chromosome abnormality, or a disorder caused by a single gene. The incidence of genetic mistakes is much higher at conception, but most of the severely defective fertilized eggs are miscarried. Genetic disorders contribute significantly to illness and death, particularly during childhood, but many of these disorders are now being treated successfully.

There are three main categories of genetic diseases – chromosome abnormalities, single-gene disorders, and disorders that result from an interaction of both genes and environment. Chromosome disorders, such as Down's syndrome, are caused by abnormalities in the number or structure of chromosomes. A few chromosome abnormalities are inherited, but the majority occur as random events at the time of conception. The risk that these sporadic genetic mistakes will happen twice in a family is usually low. Single-gene disorders are caused by a single defective gene or pair of genes. Single-gene disorders are usually inherited and the risk of recurrence in a family is often high. Defects in individual genes are sometimes the result of new mutations that occur before conception at the time an egg or sperm is formed. Sickle cell anemia, cystic fibrosis, and muscular dystrophy are among the most common and serious single-gene disorders.

Some disorders occur in people who are susceptible as a result of the interaction of several defective genes and environmental factors such as pollution and cigarette smoke. Although the inheritance of these disorders is not clear-cut, they tend to occur more frequently in relatives of affected people than in the general population. Many birth defects (such as clubfoot, spina bifida, and congenital heart disease) and many common diseases that become apparent later in life (such as diabetes and coronary heart disease) result from a combination of interrelated genetic and environmental factors.

The final section of this chapter describes the role that genes play in cancer. Research indicates that some people may inherit genes that increase their risk of cancer.

CHROMOSOME ABNORMALITIES

CHROMOSOME ABNORMALITIES include some of the most common and severe types of genetic disorders. An irregularity in the number or the structure of chromosomes can interfere in various ways with the development and functioning of a fetus. Only those fetuses with less serious chromosome abnormalities survive until birth; severely defective embryos are miscarried.

Chromosome abnormalities can result from new mutations (see page 44) or, less frequently, they can be inherited. The effects – both physical and behavioral – that chromosome abnormalities can have on people vary, depending on which chromosomes are affected and how much genetic material is involved.

About one baby in every 150 is born with some kind of chromosome abnormality. One third of those are caused by mistakes in the number of autosomes (the nonsex chromosomes), one third are caused by mistakes in the number of sex chromosomes, and one third are caused by chromosome rearrangements. Chromosome abnormalities that result from a mistake in the number of autosomes – called numerical abnormalities – usually cause multiple birth defects and mental retardation. Abnormalities of the sex chromosomes have fewer or, in some cases, no obvious effects on a person's physical or mental development.

NUMERICAL ABNORMALITIES

Mistakes can occur in a variety of ways that result in either too many or too few chromosomes in the cells of an early embryo. For example, two sperm, each with 23 chromosomes, can fertilize one egg, which also has 23 chromosomes, resulting in 69 chromosomes instead of the usual 46. An embryo with such an abnormality, however, seldom survives long after conception. Most numerical abnormalities involve only one chromosome. The absence of a complete chromosome, other than the X chromosome, will result in miscarriage. Even with a missing X chromosome, most pregnancies are miscarried. However, in some of these pregnancies, the fetus survives (see TURNER'S SYNDROME on page 67).

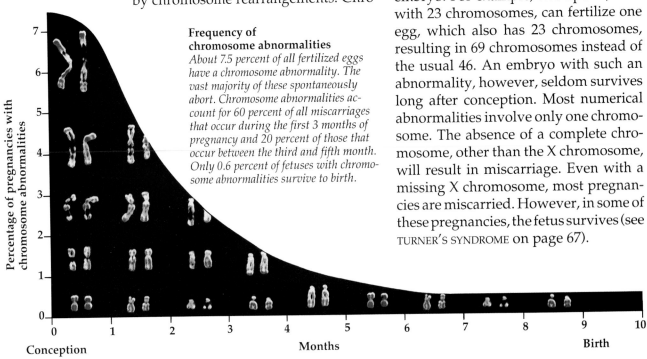

Frequency of chromosome abnormalities
About 7.5 percent of all fertilized eggs have a chromosome abnormality. The vast majority of these spontaneously abort. Chromosome abnormalities account for 60 percent of all miscarriages that occur during the first 3 months of pregnancy and 20 percent of those that occur between the third and fifth month. Only 0.6 percent of fetuses with chromosome abnormalities survive to birth.

Healthy chromosome pattern

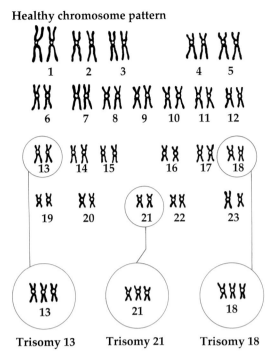

Trisomy 13   Trisomy 21   Trisomy 18

Extra chromosomes
By examining a person's chromosomes, doctors can confirm a diagnosis of a trisomy. An extra chromosome 21 causes Down's syndrome (trisomy 21). Trisomy 13 (Patau's syndrome) and trisomy 18 (Edwards' syndrome) are less common, but they cause severe defects and are usually fatal.

A few disorders occur as the result of an extra chromosome in a person's cells. All the chromosomes, except for the sex chromosomes, come in matching pairs. If a chromosome is added during cell division, a person will have three copies of that chromosome instead of the usual two. This chromosome trio is called a trisomy. Although trisomies are a frequent cause of miscarriage, one disorder – Down's syndrome (trisomy 21) – occurs relatively often in newborns.

How do numerical abnormalities occur?

Numerical chromosome abnormalities usually originate during cell division when eggs or sperm are formed (see HOW ARE SEX CELLS FORMED? on page 38). Normally, before cell division, the chromosomes of every pair in the precursor cell separate so that only one is incorporated into each egg or sperm that forms. If the chromosomes in a pair do not separate,

both may be incorporated into one of the eggs or sperm while another egg or sperm receives neither of the chromosomes from that pair. An embryo produced from an egg or sperm with such a defect will have either an extra chromosome or a missing chromosome.

The failure of chromosomes to separate during sex cell formation – called nondisjunction – occurs more frequently in older women. The incidence of Down's syndrome, in particular, has been linked to the age of the mother.

How nondisjunction works
A trisomy can occur when a pair of chromosomes do not separate during cell division and, as a result, are not equally distributed among the newly formed eggs or sperm. Sex cells with a missing chromosome do not survive.

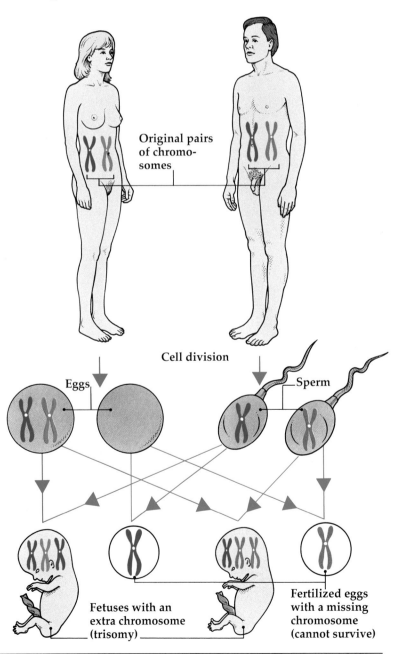

Original pairs of chromosomes

Cell division

Eggs

Sperm

Fetuses with an extra chromosome (trisomy)

Fertilized eggs with a missing chromosome (cannot survive)

CHROMOSOME ABNORMALITIES: WHAT ARE THEIR EFFECTS?

Many fetuses with chromosome abnormalities do not survive to birth. Babies who do survive may have varying degrees of mental and physical impairment or may not have any noticeable effects at all. Some chromosome disorders – called syndromes – produce a combination of recognizable physical changes. For example, trisomy 13 (Patau's syndrome) is characterized by a cleft lip and palate, extra fingers and toes, a small head, port-wine-colored birthmarks, and scalp irregularities.

ABNORMALITIES OF THE SKIN

Depigmented skin
Sporadic areas of skin that lack pigment may indicate mosaicism (see page 65), in which some cells in a person's body have a different number of chromosomes than the other cells.

Scalp defects
A baby with trisomy 13 may have irregularities of the scalp, such as ulceration.

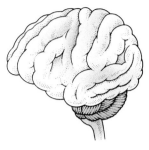

ABNORMAL BRAIN DEVELOPMENT AND MENTAL RETARDATION

FACIAL ABNORMALITIES

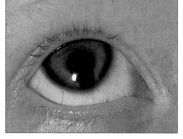

CHROMOSOME ABNORMALITY

Brushfield spots
White spots on the iris of the eye, known as Brushfield spots, are a common sign of Down's syndrome but they can occur in unaffected people as well.

Defective iris
Many eye abnormalities, such as a failure of the iris to develop fully, are associated with chromosome disorders.

INTERNAL ABNORMALITIES

HEART AND LUNG DEFECTS

GASTROINTESTINAL DEFECTS

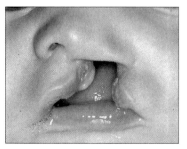

Cleft lip
Birth defects such as a cleft (split) lip and cleft palate (a gap in the roof of the mouth) can be caused by serious chromosome disorders such as trisomy 13. However, most of these defects occur independently of other genetic abnormalities.

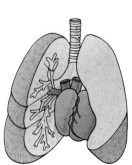

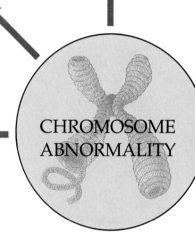

ABNORMALITIES OF THE HANDS AND FEET

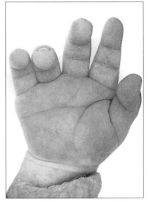

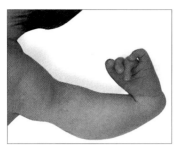

Malformed limb
Chromosome disorders including trisomy 18 can cause limb defects such as rigid joints.

RETARDED GROWTH

Single palm crease
A single crease in the palm of each hand is a common sign of Down's syndrome. A single palm crease is also occasionally found in genetically normal people.

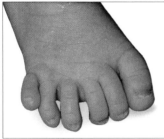

Extra toes
Extra toes and fingers can result from several chromosome disorders, including trisomy 13, but they can also occur in children who do not have abnormal chromosomes.

ABNORMAL SEXUAL DEVELOPMENT/ INFERTILITY

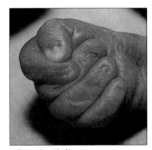

Clenched fists
A clenched hand position with overlapping or bent fingers often occurs in children with trisomy 18.

KIDNEY DEFECTS

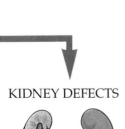

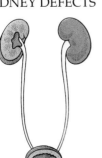
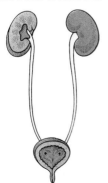

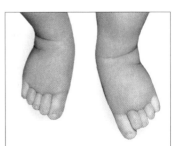
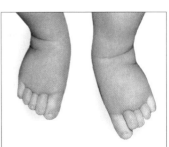

Swollen feet
A buildup of fluid in a newborn girl's feet and/or hands may be the only sign at that time that she has Turner's syndrome, which is caused by the absence of one of the two female X sex chromosomes.

CHROMOSOME TESTS

All known chromosome disorders, including Down's syndrome, can be detected early in a pregnancy. Although routine screening for abnormalities is presently not feasible, you might consider being tested if you are at increased risk because of family history, a previous child with a disorder, repeated miscarriage, or infertility.

STRUCTURAL ABNORMALITIES

Structural chromosome abnormalities occur when chromosomes do not rejoin normally after breaking apart during cell division. These mistakes often occur spontaneously during the formation of eggs and sperm. The main types of abnormalities that result – called translocations, deletions, inversions, and ring chromosomes – are illustrated at right.

The effect of a chromosome rearrangement depends on whether it is balanced or unbalanced. In a balanced rearrangement, there is no gain or loss of genetic material and the rearrangement seldom causes health problems. However, a person can transmit balanced chromosomes to children in an unbalanced form. In an unbalanced rearrangement, there is a loss or gain of chromosome material, which can distort the genetic instructions and cause illness. Once formed, a structural chromosome abnormality can be passed on to offspring. Like other chromosome abnormalities, structural defects can be detected in a fetus in early pregnancy (see CHROMOSOME ANALYSIS on page 108).

Translocations

Translocations result from the exchange of chromosome material between two different chromosomes or the fusion of two chromosomes. In a balanced translocation, the exchange of genetic material between chromosomes does not involve either loss or gain. A balanced translocation does not necessarily cause any mental or physical defects and may, in fact, go undetected throughout a person's life. That person, however, is at increased risk of having a child with a genetically unbalanced chromosome abnormality such as Down's syndrome. In unbalanced translocations, genetic material has either been added or lost, potentially causing severe abnormalities and often resulting in miscarriage.

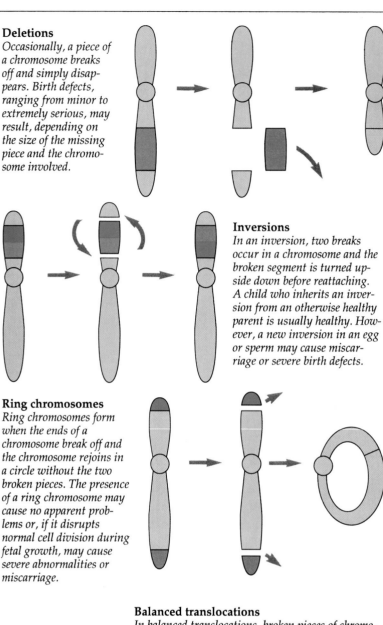

Deletions
Occasionally, a piece of a chromosome breaks off and simply disappears. Birth defects, ranging from minor to extremely serious, may result, depending on the size of the missing piece and the chromosome involved.

Inversions
In an inversion, two breaks occur in a chromosome and the broken segment is turned upside down before reattaching. A child who inherits an inversion from an otherwise healthy parent is usually healthy. However, a new inversion in an egg or sperm may cause miscarriage or severe birth defects.

Ring chromosomes
Ring chromosomes form when the ends of a chromosome break off and the chromosome rejoins in a circle without the two broken pieces. The presence of a ring chromosome may cause no apparent problems or, if it disrupts normal cell division during fetal growth, may cause severe abnormalities or miscarriage.

Balanced translocations
In balanced translocations, broken pieces of chromosomes are reattached in different combinations without a loss or gain of genetic material. A child who inherits a balanced translocation from an otherwise healthy parent is usually healthy.

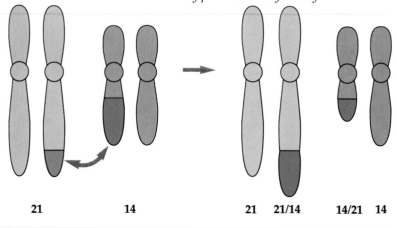

21   14   21 21/14 14/21 14

CHIMERISM

Rarely, a person may carry two different cell populations. That person is called a chimera. A chimera develops when two fertilized eggs are fused, or when a small number of cells of nonidentical twins are exchanged in the placenta. Some medical treatments can also produce chimerism. For example, after undergoing a bone marrow or organ transplant, you may have two distinct cell populations because the donor cells differ genetically from your own. Chimerism does not usually cause health problems.

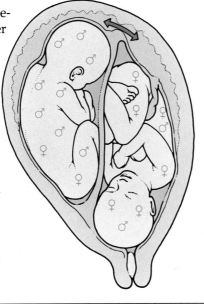

Blood chimeras
In a pregnancy involving non-identical twins, one female and one male, some cells may be exchanged through the placentas. Some of the male's blood cells have a female chromosome pattern (XX) and some of the female's cells have a male chromosome pattern (XY). The twins' health is unaffected.

DOWN'S SYNDROME

Down's syndrome is caused by the presence of an extra chromosome 21. The disorder is the single most common chromosome abnormality, occurring at a rate of about one in 800 live births.

At birth, babies with Down's syndrome are slightly smaller than average. They have loose joints, weak reflexes, and poor muscle tone. They may also have characteristic facial features.

Although all children with Down's syndrome are mentally retarded to some degree, many are capable of a limited amount of learning, including, in some cases, the ability to read. Children with Down's syndrome often have an unusually cheerful and affectionate nature.

About half of children with Down's syndrome are born with severe heart defects; many do not survive their first year. Defects of the digestive tract, kidneys, thyroid gland, and reproductive organs are also more common in children with Down's syndrome.

MOSAICISM

Mosaicism is a condition in which a person has a mixture of normal and abnormal cells. Cells with an incorrect number of chromosomes can often be detected in a blood sample. Some of the cells will have a normal chromosome pattern, while others will have either too many or too few chromosomes. Sometimes, the abnormal cells are confined to particular tissues rather than to blood cells; in these cases, a diagnosis requires a tissue examination. The effects of mosaicism depend on the proportion of abnormal cells to healthy cells in a person's body. For example, mosaic Down's syndrome may be apparent in 90 percent of a person's blood cells but in only 20 percent of his or her brain cells; this person may have almost normal intelligence. The greater the proportion of healthy cells, the healthier the person will be.

Joining in
Although children with Down's syndrome have lower IQs than other children, they can usually learn most of the basic childhood skills. With their parents' support and guidance, professional help, and the many new techniques available to encourage their development, children with Down's syndrome can achieve their full potential. These children often benefit most when allowed to attend regular schools where they learn from their peers and receive the same level of stimulation.

Children with Down's syndrome are more prone to infections, eye abnormalities, and skin disorders; for unknown reasons, they are also 20 times more likely than other children to have leukemia.

The child of a woman with Down's syndrome has a 1 in 3 risk of also having the disorder. However, people with Down's syndrome rarely reproduce and most males with the disorder are sterile.

What causes Down's syndrome?

Most cases of Down's syndrome result from nondisjunction, which is the failure of a pair of chromosomes to separate, or disjoin, during cell division when a parent's eggs or sperm are formed. This results in an egg or a sperm and all of its descendant cells having an extra chromosome. A woman's risk of having a child with Down's syndrome increases with her age. The risk that a couple will

have a second child with Down's syndrome is only slightly higher than the risk (which is also dependent on the mother's age) of the general population.

A small number (about 5 percent) of people with Down's syndrome have a chromosome rearrangement called an unbalanced translocation. Although this can occur spontaneously during egg or sperm formation when genetic material has been added or lost, about half of the time it occurs when a child inherits an abnormal set of chromosomes from a healthy parent who carries a balanced chromosome translocation. A couple's risk of having another child with Down's syndrome caused by a translocation varies, depending on the type of translocation and the chromosomes involved and

Mother's age and Down's syndrome
The risk of having a child with Down's syndrome increases with the age of the mother. For a woman under the age of 20, the risk of having a child with Down's syndrome is 1 in 1,500; at 35 it is about 1 in 400 (2.6 per 1,000 births); and by age 45 it is about 1 in 28 (35.7 per 1,000 births). The risk of any type of chromosome abnormality, including Down's syndrome, is about twice the risk of Down's syndrome alone and also increases with the mother's age.

Frequency per 1,000 births

Age	30	35	37	39	41	43	45
	1.1	2.6	4.2	6.7	11.8	20.0	35.7

on which parent is the carrier. The risk of recurrence is higher when the mother is the carrier. Because balanced translocations can be inherited, relatives of an identified carrier parent should be tested for the translocation because they may also be at increased risk of having a child with Down's syndrome.

Life expectancy

Most of the deaths related to Down's syndrome occur in the first year of life and result from heart defects. Seventy-five percent of people with Down's syndrome survive to age 20, and 25 percent live to age 50. Many older people with Down's syndrome have dementia similar to Alzheimer's disease.

SEX CHROMOSOME ABNORMALITIES

Abnormalities in the number of sex chromosomes are relatively common, occurring in about one in every 350 males and one in every 550 females. Most are caused by mistakes that occur during the formation of eggs and sperm when two sex chromosomes fail to separate during cell division. These abnormalities rarely occur twice in a family.

Turner's syndrome

In Turner's syndrome, one of the sex chromosomes has been lost, producing a female with only 45 chromosomes instead of the usual 46. Although Turner's syndrome is the cause of about 20 percent of all miscarriages, the fetuses that survive are usually born healthy and will have normal intelligence. However, some girls with Turner's syndrome are born with heart disease caused by a narrowing of the main artery from the heart. At birth, many babies with Turner's syndrome can be identified by a swelling of their hands and feet and characteristic folds of skin on the back of their neck. Short stature and infertility are the most common features of Turner's syndrome.

Females with an extra X chromosome

One in 1,000 females has an extra X sex chromosome, a condition called triple X syndrome. Although they have no apparent physical abnormalities, females with triple X syndrome may have slightly lower than average intelligence. About half of them have delayed speech development. Women with triple X syndrome are fertile but they may go through menopause at an early age. Their children usually have normal chromosomes.

FRAGILE X SYNDROME

Fragile X syndrome is the most common inherited form of mental retardation. The disorder is linked to a pinched-in area on the X chromosome that looks ready to break. Women who carry the genetic defect may be unaffected but most men with the disorder have a degree of mental impairment, ranging from mild to severe, and often have distinctive facial features. A new test that analyzes genetic material can now determine whether a fetus or newborn carries the fragile X gene. The disorder has a high risk of recurrence in families.

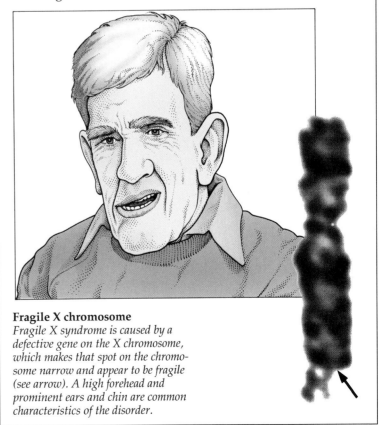

Fragile X chromosome
Fragile X syndrome is caused by a defective gene on the X chromosome, which makes that spot on the chromosome narrow and appear to be fragile (see arrow). A high forehead and prominent ears and chin are common characteristics of the disorder.

CASE HISTORY
RISK OF DOWN'S SYNDROME

SUSAN HAS TWO TEENAGE children from her first marriage. She remarried after her husband's death, and, when she was 38, she and her second husband decided to have a child. Because of Susan's age, they were concerned about the risk of having a child with Down's syndrome. They decided to seek advice from Susan's doctor.

PERSONAL DETAILS
Name Susan Morris
Age 38
Occupation Secretary
Family There are no serious health problems in the family histories of Susan or her husband. Both of her children are healthy.

MEDICAL BACKGROUND
Susan and her husband are healthy. She has had two normal pregnancies and one early miscarriage.

THE CONSULTATION
Susan expresses her concerns about having a baby with Down's syndrome. The doctor questions her about her family's medical background and uses the information to draw a diagram of the family's medical history. The history does not reveal any disorders that are likely to pose a risk to a future pregnancy. The doctor explains that early miscarriage is common and that Susan's miscarriage years ago does not increase the risk of a chromosome abnormality in a future pregnancy. Susan is told that, if she has a baby at the age of 39, the risk that the baby will have Down's syndrome is about 1 in 150 and that, by age 43, the risk will increase to about 1 in 50.

The doctor tells Susan and her husband that a fetus can be tested for Down's syndrome in the early part of pregnancy. He explains, however, that the tests cannot detect non-chromosome-related birth defects. Susan decides that she wants to have a prenatal chromosome test. The doctor explains that one of the tests – chorionic villus sampling – can be performed at an earlier stage of pregnancy, allowing for safer termination of a pregnancy if that is her choice. However, Susan decides to have a different prenatal test called amniocentesis because she is told that studies have shown that it carries a lower risk of miscarriage.

FURTHER INVESTIGATION
Susan becomes pregnant 4 months after the consultation. The pregnancy progresses normally and her obstetrician performs amniocentesis during the 16th week. Using an ultrasound image for guidance, the obstetrician inserts a needle into a fluid-filled area of the uterus and withdraws a sample of clear amniotic fluid for laboratory analysis. The amniotic cells from the sample are grown in the laboratory and, after 2 weeks, they are examined under a microscope. The fetus is found to have a normal chromosome pattern. The couple decided that they also want to know the sex of the fetus; the results of the amniocentesis show that the fetus is male.

THE OUTCOME
Susan and her husband are relieved that their baby does not have a chromosome disorder. Susan's pregnancy continues without any complications and, at 39 weeks, a healthy baby boy is born.

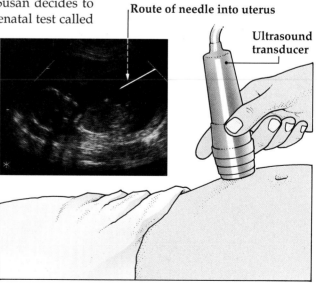

Route of needle into uterus

Ultrasound transducer

Amniocentesis
To take a sample of amniotic fluid, the doctor inserts a needle through Susan's abdominal wall and uterus under the guidance of ultrasound. The ultrasound scanner allows the doctor to see the exact position of the needle and the fetus inside the uterus.

Males with an extra X chromosome

One in 1,000 males has an extra X sex chromosome, giving him an XXY sex chromosome set instead of the usual XY. The condition is called Klinefelter's syndrome. Because their testicles do not produce sperm, XXY males are sterile, but they are able to have sex. Treatment with male hormones can sometimes help them progress through puberty. Some boys with Klinefelter's syndrome have educational and behavioral problems, but many function well socially.

What gives males an extra X?
Klinefelter's syndrome can originate at conception with the union of a normal sperm and an abnormal egg (which has two X chromosomes instead of one), or an abnormal sperm (with both an X and a Y chromosome instead of one or the other) and a normal egg.

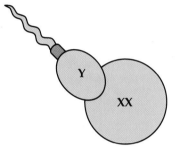

Normal sperm and abnormal egg

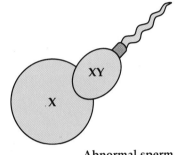

Abnormal sperm and normal egg

THE MYTH OF THE XYY MALE

About one in every 1,000 males is born with an extra Y chromosome, giving him the sex chromosome set XYY instead of XY. While this chromosome pattern has been linked to an increased risk of incarceration for criminal behavior, studies show that most XYY males are no more violent or aggressive than other men. XYY males often have lower IQs than their siblings and sometimes have learning disabilities and behavioral problems.

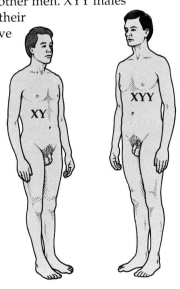

XYY syndrome
Men who carry an extra Y chromosome have no major physical problems but tend to be taller than average.

ASK YOUR DOCTOR
CHROMOSOME ABNORMALITIES

Q My brother has Down's syndrome. Am I at increased risk of having a child with Down's?

A Yes, your risk is increased, especially if your brother has an uncommon type of Down's syndrome caused by a rearrangement of pieces from two different chromosomes, called a translocation, that he inherited from one of your parents. A chromosome analysis of your brother's blood can determine the cause of his Down's syndrome.

Q My first pregnancy ended in an early miscarriage. Because my mother had three miscarriages, I am worried that I may have another one. Could there be a genetic cause?

A Miscarriage is extremely common and does not necessarily result from a genetic disorder. Recurrent miscarriages, however, may be associated with chromosome translocations, which can be inherited. A chromosome analysis of your blood can confirm whether or not you have a balanced translocation.

Q Our first child was born with a heart defect. Can chromosome tests determine our chances of having another affected child?

A Congenital heart defects, except when they are accompanied by other birth defects, are not caused by chromosome abnormalities. A normal result on a chromosome test would neither verify nor exclude a genetic risk to your children. Although greater than the risk in the general population, your risk of having another child with a heart defect is relatively low.

SINGLE-GENE DISORDERS

GENETICISTS HAVE SO FAR identified about 4,000 different genetic disorders that are caused by single abnormal genes. The disease-causing gene may be located on one of the nonsex chromosomes (autosomes) or on one of the sex chromosomes (almost always the X sex chromosome). Single-gene disorders can be inherited in a number of ways.

The nonsex chromosomes, or autosomes, come in 22 pairs. Most disorders caused by defects in single genes on autosomes affect both sexes equally. An autosomal disorder is classified as autosomal dominant (see page 71) or autosomal recessive (see page 75), depending on whether the disease-causing gene is dominant or recessive. If the gene is recessive, two copies are required to cause illness.

Most sex-linked disorders are caused by a gene located on the X chromosome. X-linked recessive disorders (see page 80) usually affect males who inherit the disease gene from their mothers, who carry the gene but are unaffected. X-linked dominant disorders, which affect more females than males, are rare.

INHERITANCE PATTERNS

Individually, most single-gene disorders are rare but, taken together, they account for significant illness and disability.

Single-gene disorders are inherited in the same ways that physical characteristics such as eye color and nose shape are; their inheritance patterns are usually easy to predict. Some disorders (such as Huntington's chorea) are inherited in an autosomal dominant pattern, some (such as cystic fibrosis) are inherited in an autosomal recessive pattern, and some (such as Duchenne type muscular dystrophy and hemophilia) follow an X-linked recessive pattern.

Different genes sometimes cause the same disorder. As a result, the inheritance pattern of a disorder may vary greatly among families. For example, the various genes that cause retinitis pigmentosa (degeneration of the retinas of both eyes) can be autosomal dominant, autosomal recessive, or X-linked recessive. When only one family member is affected by such a disorder, it is difficult to determine the inheritance pattern.

One affected parent and one healthy parent

Healthy carrier parents

Affected child (1 in 2 risk)

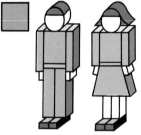

Affected child (1 in 4 risk)

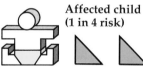

Healthy father **Healthy carrier mother**

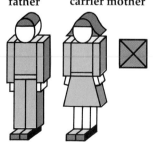

Affected male child (1 in 2 risk for boys; girls are unaffected)

 Abnormal autosomal dominant gene

 Abnormal autosomal recessive gene

 Abnormal X-linked recessive gene

How are single-gene disorders inherited?
The inheritance patterns of single-gene disorders and their risks are shown here. Dominant disorders are inherited from one affected parent; recessive disorders are inherited from unaffected parents who both carry the abnormal gene; X-linked recessive disorders are inherited by boys from mothers who carry the gene but are themselves unaffected.

INBORN ERRORS OF METABOLISM

Disorders that result from defects in genes that control the body's chemical functioning (metabolism) are referred to as inborn errors of metabolism. The connection between genes and a malfunction in metabolism was first made in 1902 by the English doctor Archibald Garrod. He proposed that all metabolic processes work in steps and that each step is controlled by a specific enzyme, which in turn is the product of a particular gene. If a gene is defective, the enzyme it makes will be defective or in short supply. A defective or missing enzyme can block an essential step in the metabolic process and cause illness. Most of the approximately 200 inborn errors of metabolism known today are inherited in an autosomal recessive or X-linked recessive pattern.

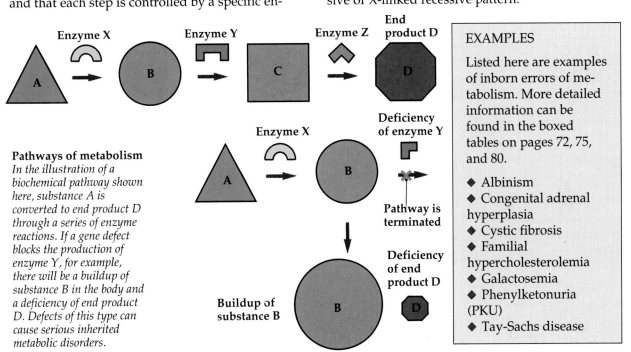

Pathways of metabolism
In the illustration of a biochemical pathway shown here, substance A is converted to end product D through a series of enzyme reactions. If a gene defect blocks the production of enzyme Y, for example, there will be a buildup of substance B in the body and a deficiency of end product D. Defects of this type can cause serious inherited metabolic disorders.

EXAMPLES

Listed here are examples of inborn errors of metabolism. More detailed information can be found in the boxed tables on pages 72, 75, and 80.

- ◆ Albinism
- ◆ Congenital adrenal hyperplasia
- ◆ Cystic fibrosis
- ◆ Familial hypercholesterolemia
- ◆ Galactosemia
- ◆ Phenylketonuria (PKU)
- ◆ Tay-Sachs disease

AUTOSOMAL DOMINANT DISORDERS

Autosomal dominant disorders are caused by changes, or mutations, in dominant genes carried on autosomes (the nonsex chromosomes). These disorders affect the sexes equally. Only one copy of a dominant gene is needed to have an effect on a person. Some of these disorders, such as hereditary spherocytosis (a red blood cell disorder that leads to anemia), are relatively mild and can be treated. Others, such as Huntington's chorea, are currently untreatable and fatal. Many autosomal dominant disorders, including Huntington's chorea, do not manifest themselves until a person reaches adulthood.

Pattern of inheritance
The family tree below, which shows the inheritance of achondroplasia (a type of dwarfism) illustrates typical autosomal dominant inheritance. People who have the defective dominant gene exhibit the disorder. The effects of achondroplasia vary little among those with the gene.

■ **Affected male**   □ **Unaffected male**
● **Affected female**   ○ **Unaffected female**

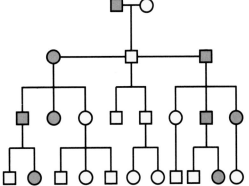

Achondroplasia
People with achondroplasia are short with disproportionately short limbs. They have a well-developed body and a large head with a prominent forehead. They are of normal intelligence.

Inheritance

Autosomal dominant disorders affect males and females equally. A person who has a dominant disease-causing gene will almost always be affected by it. Each of that person's children has a 50-50 chance of inheriting the gene and, therefore, of having the disorder.

It is rare for both parents to have the same dominant disorder. If they do, each of their children has a 25 percent chance of inheriting the gene from both of them. A double dose of a dominant disease-causing gene produces a much more severe effect than does a single dose. A person is more likely to get a double dose of a gene that occurs frequently in the population. Because one in 500 people has the dominant gene for familial hypercholesterolemia (which causes high cholesterol levels), it is one of the most widespread genetic disorders. Each of the children of two people with the gene has a 25 percent chance of inheriting the gene from both of them and of having severe heart disease at a young age.

EXAMPLES OF AUTOSOMAL DOMINANT DISORDERS	
DISORDER	**EFFECTS**
Acute intermittent porphyria	Buildup of substances (called porphyrins), which affects the nervous system and other organs, causing abdominal pain, vomiting, and arm and leg cramps
Adult polycystic kidney disease	Presence of numerous cysts in the kidneys, eventually leading to kidney failure
Familial hyper-cholesterolemia	High blood cholesterol levels and early coronary heart disease
Familial polyposis	Development of many polyps in the colon and rectum that can lead to cancer
Huntington's chorea	Degeneration of nerve cells in the brain, leading to jerky, involuntary movements and dementia
Marfan's syndrome	Abnormality of connective tissue formation, resulting in unusually long limbs; also heart and eye defects
Neurofibromatosis	Development of patches of pigmented skin and many soft, fibrous swellings that grow from nerves in the skin
Tuberous sclerosis	Seizures, mental retardation, muscle tumors, tumors behind the eye, and an acnelike rash on the face

Acute intermittent porphyria

The diagram below illustrates the inheritance pattern in part of the British royal family of the autosomal dominant disorder acute intermittent porphyria. Geneticists believe that the metabolic disorder may have affected the mental state and judgment of George III (1738-1820) during his reign.

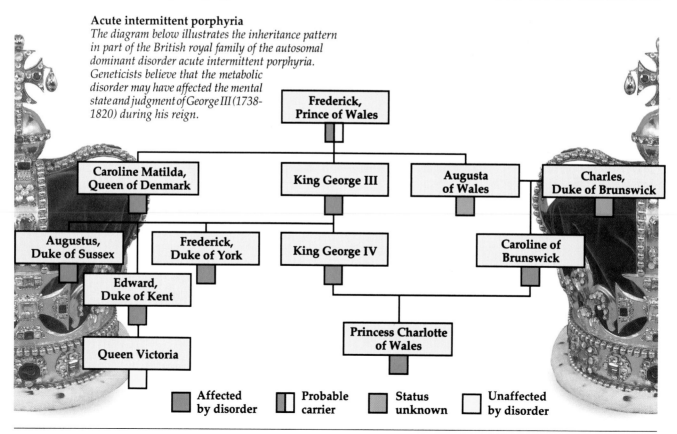

Affected by disorder   Probable carrier   Status unknown   Unaffected by disorder

CASE HISTORY
A NEW GENE MUTATION

LISA AND ROBERT **have been married for 3 years and want to start a family. However, one of Lisa's sisters has a daughter with a type of dwarfism called achondroplasia and she is worried that the disorder may run in her family. Before Lisa gets pregnant, the couple decides to seek advice from their doctor.**

PERSONAL DETAILS
Name Lisa Carlucci
Age 28
Occupation Bartender
Family Lisa's niece has achondroplasia.

MEDICAL BACKGROUND
Neither Lisa nor her husband has any health problems.

THE CONSULTATION
The couple tell the doctor that they are worried about how they would cope with a child with a mental or physical handicap. Their fears arose when Lisa's sister's baby was born with a type of dwarfism called achondroplasia. The doctor tells the couple that, although achondroplasia is caused by an abnormality in a single gene, it is not always acquired through inheritance. The doctor explains that, because Lisa's sister and her husband exhibit none of the signs of the dominant disorder, such as disproportionately short limbs, neither one has the defective gene – anyone who has the gene for achondroplasia is affected by it. It is obvious, therefore, that the condition occurred in their daughter as the result of a new gene mutation. Lisa's risk of having a baby with achondroplasia is no greater than that of the general population.

The doctor says that genetic abnormalities can occur unexpectedly in any pregnancy and that about 6 to 8 percent of all babies are born with a birth defect or genetic disorder of varying severity. She tells the couple that ultrasound scans can be used during Lisa's pregnancy to monitor the fetus's development and check for major structural abnormalities, including achondroplasia.

FURTHER INVESTIGATION
Although she understands that her risk of transmitting a genetic disorder to her children is no greater than average, Lisa is not completely reassured. When she becomes pregnant later that year, she asks her obstetrician to arrange for her to have an ultrasound scan. The ultrasound, which is given during the 18th week of pregnancy, shows that the fetus is developing normally. Lisa is especially happy to learn that the length of the fetus's thighbone is normal for this stage of pregnancy, excluding the possibility of achondroplasia.

THE OUTCOME
Lisa gives birth to a healthy girl. The doctor finds no evidence of achondroplasia or any other abnormalities in the baby.

Checking the family history
During her consultation with Lisa and Robert, the doctor studies their family history and finds no disorders that are likely to increase the risk of genetic abnormalities in their children.

Variations in gene activity

Many disease-causing dominant genes can affect people to varying degrees. Occasionally, a defective dominant gene will not produce any effect at all in a person; geneticists do not understand why this happens. Even in the same family, one person with a dominant disease gene may have no symptoms of illness, while another is disabled by it. However, the effects of dominant genes usually lie somewhere between the two extremes. Although a person may have a defective dominant gene and be unaffected by it, his or her child can inherit the gene and be severely affected.

New mutations

Autosomal dominant disorders can suddenly appear in a family when a new mutation occurs during the formation of a parent's egg or sperm. New gene mutations are easy to identify because both parents are obviously unaffected. Most cases of achondroplasia are caused by new mutations. Other genetic disorders, such as tuberous sclerosis, result from new mutations some of the time. However, because these disorders can have varying effects, it may be difficult to determine if a parent is affected and, therefore, whether a child's disorder was

Marfan's syndrome
Abraham Lincoln's lanky body, drifting eye, and other physical peculiarities have led some doctors to believe that he had the autosomal dominant disorder Marfan's syndrome. The disorder is characterized by long legs, arms, and fingers, irregularities in the position of the lens of the eye, and heart and blood vessel abnormalities.

SEX-CELL MOSAICISM

Occasionally, an autosomal dominant gene mutation is present in only some of a person's cells (a phenomenon called mosaicism). If those cells are confined to the testicles or ovaries, the person has no symptoms of the disorder but can transmit it to children through a sperm or egg that carries the mutation. Because several sperm or eggs may carry the mutation, more than one child can be affected.

inherited or the result of a new mutation. For several dominant disorders, including Marfan's syndrome, the risk of a new mutation, while still relatively low, increases with the age of the father. Other dominant disorders, such as Huntington's chorea, are seldom caused by new mutations.

What can be done?

Some dominant disorders can be treated. In the case of a severe, untreatable disorder, couples can be counseled about their risk of having an affected child and their options for avoiding those risks. After a careful physical examination, a doctor can sometimes determine if a healthy person with a family history of a disorder carries the disease-causing gene. It is often difficult to evaluate the risks if the effects of a disorder vary considerably, if the disorder is not always exhibited in people who have the gene, or if the disorder does not appear until later in life. In such cases, it may be possible to determine the risks more accurately using DNA analysis (see page 109), which is done for disorders such as Huntington's chorea and Marfan's syndrome.

Tuberous sclerosis
Tuberous sclerosis, a disorder of the nervous system, is a good example of an autosomal dominant disorder that can have widely varying effects on people. For example, a parent with only a skin manifestation of the disorder (right) may have a child with severe symptoms, including seizures and mental retardation.

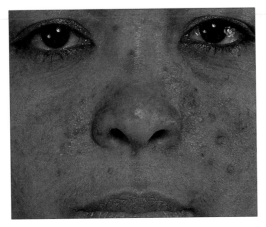

AUTOSOMAL RECESSIVE DISORDERS

Many genetic disorders – ranging from extremely rare diseases to those that occur frequently in some populations – follow a pattern of autosomal recessive inheritance. For example, cystic fibrosis (see page 76) is the most common autosomal recessive disorder among white Americans of European descent (one in 20 carries the recessive gene). The disease strikes one in 2,000 of their children. About one in 10 black Americans carries the recessive gene for sickle cell anemia; nearly one in 500 black American children is born with the disease. Among

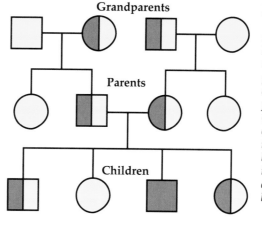

Pattern of inheritance
The diagram at left shows the typical pattern of inheritance of an autosomal recessive disorder. The parents are carriers who each received the gene from one of their parents (grandparents here). Each child has a 1 in 4 chance of inheriting the gene from both parents and of having the disorder; an unaffected child has a 2 in 3 chance of being a carrier.

Affected male / Unaffected male / Healthy carrier (male) / Healthy carrier (female) / Unaffected female

EXAMPLES OF AUTOSOMAL RECESSIVE DISORDERS

DISORDER	EFFECTS
Albinism	Lack of pigment in hair, skin, and eyes, which leads to extreme skin sensitivity and visual defects; caused by an enzyme deficiency
Beta thalassemia *	Anemia and irregular bone growth; caused by deficiency of hemoglobin in red blood cells
Congenital adrenal hyperplasia †	Masculinization of female genitals if untreated; caused by an enzyme deficiency
Congenital deafness (some forms)	Complete inability to hear
Cystic fibrosis *	Chronic lung infections and an inability to absorb fats and other nutrients
Friedreich's ataxia	Degeneration of nerve fibers in the spinal cord, leading to loss of coordination
Galactosemia †	Cataracts, mental retardation, and cirrhosis if untreated; caused by an enzyme deficiency
Hemochromatosis	Storage of too much iron, leading to diabetes, cirrhosis of the liver, heart disease, and tissue damage
Phenylketonuria (PKU) * †	Mental retardation if untreated; caused by a lack of the liver enzyme phenylalanine hydroxylase
Sickle cell anemia * †	Anemia and attacks of severe bone and abdominal pain; caused by production of abnormal red blood cells
Tay-Sachs disease *	Fatal buildup of a harmful substance in the brain; caused by an enzyme deficiency

* Carrier testing available   † Newborn testing available

Italian and Greek Americans, one in 10 is a carrier of the gene for the blood disease beta thalassemia; one in 400 of their children is born with the disorder. Some examples of autosomal recessive disorders are listed in the table at left.

Inheritance

Autosomal recessive disorders affect people who have two defective copies of a recessive gene. A carrier of an autosomal recessive disorder has a normal copy of the gene as well as a defective copy. Because the normal gene can make enough of the required protein on its own, a carrier is usually not affected. Each of his or her children has a 50 percent chance of inheriting the abnormal gene and of also being a carrier. If both parents are carriers, each child has a 25 percent chance of inheriting two copies of the gene and having the disorder.

Most of us carry several abnormal recessive genes. But, because most of them are rare, the chance of any two unrelated people having the same defective gene is small. Because relatives are more likely to share a rare recessive gene, they are more likely to produce offspring with a rare genetic disorder such as congenital adrenal hyperplasia (see MARRIAGES BETWEEN RELATIVES on page 78).

WHAT IS CYSTIC FIBROSIS?

Cystic fibrosis, the major genetic killer of white children, is a single-gene disorder with two main features – chronic lung infections and insufficient absorption of nutrients by the digestive tract – that lead to poor growth and development. The disorder causes a buildup of salt in sweat and thick mucus in the lungs and pancreas. The symptoms of cystic fibrosis, which can develop from birth on, eventually lead to degeneration of the lungs.

Signs and symptoms
The symptoms of cystic fibrosis and their severity can vary widely among people with the disorder. Although the disease is present at birth, it may not be detected for months or even years. However, a person with cystic fibrosis usually has some or all of the signs and symptoms shown here.

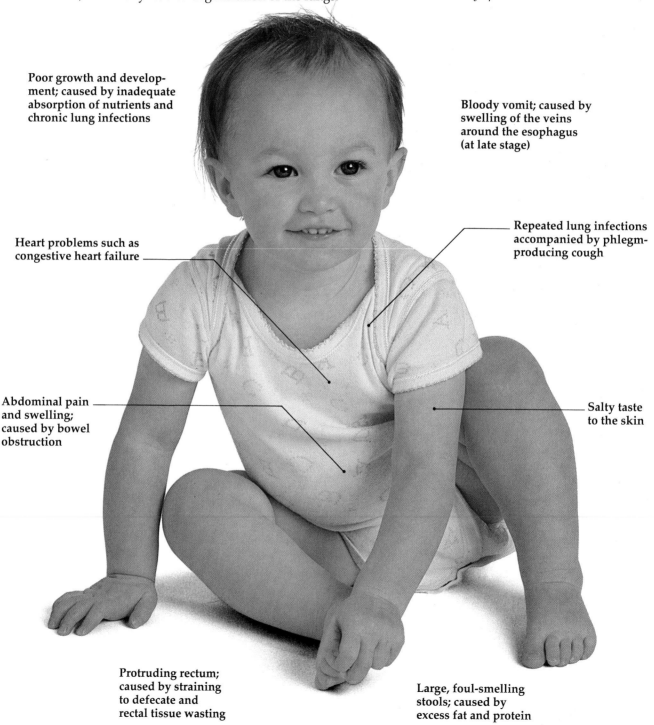

Poor growth and development; caused by inadequate absorption of nutrients and chronic lung infections

Bloody vomit; caused by swelling of the veins around the esophagus (at late stage)

Heart problems such as congestive heart failure

Repeated lung infections accompanied by phlegm-producing cough

Abdominal pain and swelling; caused by bowel obstruction

Salty taste to the skin

Protruding rectum; caused by straining to defecate and rectal tissue wasting

Large, foul-smelling stools; caused by excess fat and protein

Pattern of inheritance

Cystic fibrosis follows an autosomal recessive pattern of inheritance. It is the most common autosomal recessive disorder among whites, especially those of northern European descent. It affects one in 2,000 live births in this group. About one in 20 whites is a carrier.

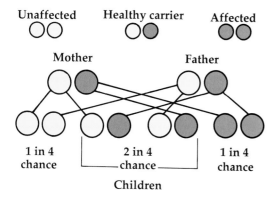

Unaffected   Healthy carrier   Affected

Mother   Father

1 in 4 chance   2 in 4 chance   1 in 4 chance

Children

Genetic counseling

The parents of a child with cystic fibrosis both have a copy of the recessive disease-causing gene and are therefore carriers. Each of their children has a 1 in 4 chance of having the disorder; each of their unaffected children has a 2 in 3 chance of being a carrier of the gene. Genetic counseling is available for anyone concerned about being a carrier of a serious recessive disorder.

Diagnosing cystic fibrosis

If your child has suspicious symptoms – particularly repeated lung infections and a failure to grow – you should inform your doctor immediately. Early treatment is essential to minimize lung damage. A high salt content in your child's sweat and abnormal levels of proteins in the blood are signs of the disorder. A chest X-ray might show scattered areas of lung infection – shown below are a normal lung (left) and a lung affected by cystic fibrosis (right).

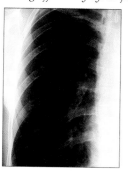
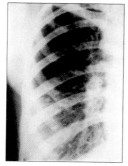

Discovering the gene

The primary gene for cystic fibrosis has recently been located and identified by DNA analysis (left). This breakthrough has made it possible to detect carriers among the relatives of affected children. Prenatal diagnosis is available for parents who have one child with the disease and are contemplating having another child, and couples who both have been found to be carriers (see page 82). The discovery of the cystic fibrosis gene has led to hopes for a cure by means of gene therapy (see page 135).

Management

There is currently no cure for cystic fibrosis. However, the severity of the disorder can be reduced dramatically with the use of antibiotics. Physical therapy can help a person cough up secretions from his or her lungs. Other treatment includes taking enzyme preparations with meals to help digestion.

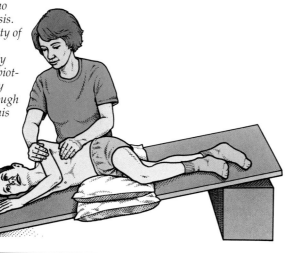

Outlook

The highly specialized treatment available to people with cystic fibrosis gives them a much better quality of life than was possible before; more than two thirds now survive to adulthood. New genetically engineered drugs hold the promise, for the first time, of controlling the disease. Breathed into the lungs in an aerosol spray, the drugs loosen mucus, improving lung capacity.

Screening and treatment

Before a couple has had a child with an autosomal recessive disorder, there is seldom an indication that they are carriers – most are healthy and have no family history of the disorder. For some common recessive disorders that occur in particular families or ethnic or racial groups, people can now be tested to find out if they are carriers. For many autosomal recessive disorders, a prenatal diagnosis can be made using biochemical tests (which look for specific chemicals) or DNA analysis (see GENETIC ANALYSIS on page 108). Some recessive disorders, particularly metabolic disorders (see page 71) such as phenylketonuria, can be treated effectively with a restricted diet. For a few disorders, a bone marrow transplant can provide a cure. Gene therapy holds the promise of treating and curing many genetic diseases by supplementing or replacing defective genes with normal ones.

Galactosemia
Galactosemia, a metabolic disorder caused by an enzyme deficiency, follows an autosomal recessive pattern of inheritance. If untreated, the disorder can result in the development of cataracts in the eye (below) early in life.

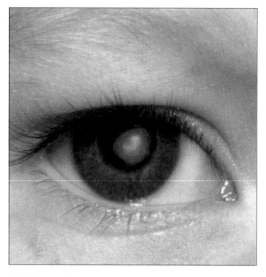

Thalassemia
Thalassemia, the collective name for a group of blood disorders, is the most common autosomal recessive disorder worldwide. In one extremely serious form – beta thalassemia – severe anemia occurs, the liver and spleen are enlarged, and skull deformities develop. In the X-ray below, the "hair-on-end" look of the edge of the skull results when the bone expands to make more room for blood-cell producing bone marrow. The skull in the inset is normal.

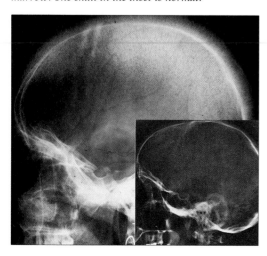

MARRIAGES BETWEEN RELATIVES

Some marriages between blood relatives are universally illegal and considered taboo, such as those between brother and sister. Others are accepted in many parts of the world. In the US, marriages between first cousins account for one in every 1,000 marriages. It is estimated that each of us carries several defective recessive genes. Because of a common ancestry, first cousins are more likely to have a rare abnormal gene in common than are two unrelated people. For this reason, their children are at a higher-than-average risk of inheriting a rare genetic disorder, such as the metabolic disorder phenylketonuria. However, the vast majority (94 percent) of children born to first cousins have no genetic defects.

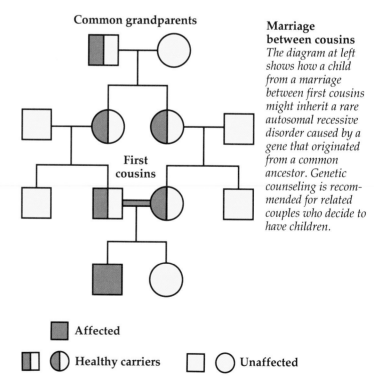

Common grandparents

First cousins

■ **Affected**

▯ ◖ **Healthy carriers**   □ ○ **Unaffected**

Marriage between cousins
The diagram at left shows how a child from a marriage between first cousins might inherit a rare autosomal recessive disorder caused by a gene that originated from a common ancestor. Genetic counseling is recommended for related couples who decide to have children.

CASE HISTORY
A LACK OF COORDINATION

JOHN IS OUTGOING and friendly and has always been a good student. However, over the past 6 months he has been moody and behaving strangely, and his schoolwork has deteriorated. His parents were especially alarmed when he became clumsy and developed a tremor in his hands.

PERSONAL DETAILS
Name John Williams
Age 13
Occupation Student
Family Both parents and John's sister are healthy.

MEDICAL BACKGROUND
John has been healthy from birth.

THE CONSULTATION
After discussing John's medical history with his parents, the family doctor examines him. The doctor notices that John is having difficulty with his coordination. The doctor also notices an unusual ring of pigmentation around the cornea of each of John's eyes. This golden-brown ring, along with John's other symptoms, indicates an abnormality in his body's ability to metabolize copper. The doctor refers John to a neurologist for further evaluation.

FURTHER INVESTIGATION
The neurologist examines John and says that he is fortunate that his family doctor realized the seriousness of his symptoms. The neurologist takes a sample of John's blood and urine and recommends that he have a liver biopsy. For this test, he refers John to a gastroenterologist.

DIAGNOSIS AND TREATMENT
The results of John's tests confirm the suspicions of the doctors. The neurologist tells John's parents that John has WILSON'S DISEASE, a rare inherited disorder that causes harmful levels of copper to accumulate in the liver. The excess copper is slowly released into the bloodstream and carried to other parts of the body, including the brain, where it collects and can cause damage. The neurologist explains that John must take a drug called penicillamine for the rest of his life. The drug increases the amount of copper excreted in the urine. The doctor tells John that the drug will maintain a safe level of copper in his body and will prevent the serious complications, such as liver and brain damage, that can result from the disorder.

GENETIC COUNSELING
The neurologist explains to John's parents that, because Wilson's disease is caused by a recessive gene, John must have inherited one copy of the gene from each of them. Because they both carry the gene, each of their children has a 1 in 4 risk of having the disorder. John's parents are not planning to have any more children but they agree that John's younger sister should be tested to determine if she has the disorder or is a carrier of the recessive gene. Fortunately, the test results indicate that she does not have the disorder, nor is she a carrier.

THE OUTCOME
Six months after starting his treatment, John is no longer having problems in school or with his coordination. He will have to continue to take penicillamine for the rest of his life and to have regular blood tests to monitor the drug's effectiveness.

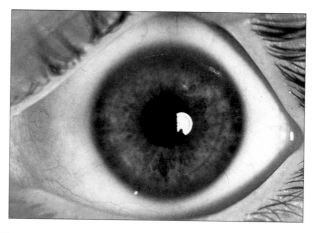

A ring of pigment
When the family doctor examines John's eyes, he finds a ring of golden-brown pigment around the outer edge of each cornea. This feature, called a Kayser-Fleischer ring, is characteristic of Wilson's disease.

X-LINKED RECESSIVE DISORDERS

All recognized sex-linked disorders are caused by defective genes located on the X chromosome. Most of these disease-causing genes are recessive. Because females have two X chromosomes, a defective recessive gene on one is usually neutralized by a normal gene on the other X chromosome. However, males have only one X chromosome – they have no matching normal gene to neutralize the effect of an X-linked disease gene.

Inheritance

Women are rarely affected by X-linked recessive disorders, but they are often carriers of the genes that cause them. In fact, a male usually inherits an X-linked disorder from his mother. Each son of a woman who carries an X-linked recessive disease gene has a 50-50 chance of

EXAMPLES OF X-LINKED RECESSIVE DISORDERS

DISORDER	EFFECTS
Duchenne type muscular dystrophy and Becker's muscular dystrophy	Progressive degeneration of muscle (Duchenne type muscular dystrophy occurs before age 6; Becker's muscular dystrophy starts later and progresses more slowly)
Fabry's disease	Kidney and blood vessel problems leading to high blood pressure, kidney failure, and stroke; skin lesions; opaque corneas in the eyes; chronic pain; and growth retardation
G6PD deficiency	Anemia that is triggered only when an affected person ingests specific drugs or foods, or during infections
Hemophilia A and B	Prolonged and severe bleeding; caused by a deficiency of either of the blood-clotting proteins factor VIII or factor IX
Hunter's syndrome	Bone deformities, progressive mental retardation, and organ damage; caused by a missing enzyme
Lesch-Nyhan syndrome	Involuntary movements, severe mental retardation, and self-mutilation; caused by a missing enzyme

inheriting the gene and having the disease. Each daughter also has a 50-50 chance of inheriting the gene, but she will be an unaffected carrier.

If a male with an X-linked disorder has children, he will pass on his affected X chromosome to all of his daughters but to none of his sons (who receive only his Y chromosome). Because the daughters who receive an X-linked recessive disease gene have a normal copy of the gene on their other X chromosome, they will be unaffected carriers. An X-linked recessive disease gene can be transmitted through many generations of a family by healthy

Female carrier

Unaffected female

Affected male

Unaffected male

Studying family inheritance patterns
An X-linked recessive disorder may appear to skip generations in a family. Study of family inheritance patterns shows that males inherit X-linked disorders from their mothers, who are healthy carriers. This pattern is illustrated above – no third-generation children have the disorder, but two females in that generation are carriers.

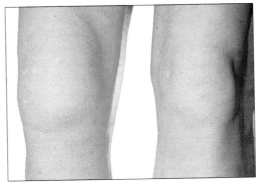

Hemophilia
The photograph shows the swollen right knee of a person with hemophilia, a disorder characterized by prolonged and severe bleeding. The swelling is caused by bleeding into the joints, which can occur as the result of even a minor injury.

Muscular dystrophy
Duchenne type muscular dystrophy is a rare X-linked recessive muscle disorder marked by a slow but progressive degeneration of muscle tissue. Affected children are usually unable to walk after about age 12. Although there is no cure, physical therapy and specialized equipment contribute considerably to quality of life.

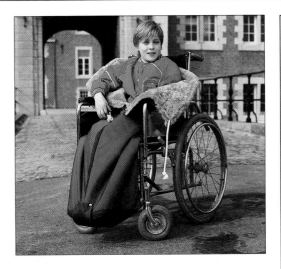

RACIAL VARIATION

Some genetic disorders occur at different rates among different racial groups. One such disorder – G6PD deficiency – is an X-linked recessive disorder caused by the lack of an enzyme that affects the chemistry of red blood cells. The disorder is more common in dark-skinned people; nearly 15 percent of black American males are affected. Symptoms of the disorder are triggered when a person with the defective gene ingests foods or drugs such as fava beans and some antimalarial drugs. Symptoms include anemia, jaundice, fatigue, headache, and shortness of breath.

carrier mothers before it becomes apparent when a boy is born with the disease for the first time in the family.

New mutations

Many of the more serious X-linked recessive disorders, such as Duchenne type muscular dystrophy, usually kill affected males before they can reproduce. These diseases would eventually become extinct if not for the fact that the abnormal gene is continually being produced spontaneously by new mutations during egg or sperm formation. There is no family history of a disorder caused by a new mutation.

X-linked recessive disorders can also arise from sex-cell mosaicism (see page 74), in which only a small percentage of a woman's eggs carry the mutation. As a result, she will not appear to be a carrier when tested but she may have more than one son with the disorder.

What can be done?

The severity of X-linked recessive disorders varies greatly. For those that are less severe, treatment is aimed at preventing or minimizing symptoms. Some disorders, such as hemophilia, are treated by regularly replacing a missing protein. Many inherited disorders will one day be cured by gene therapy, in which people will be given normal genes to do the work of defective ones (see page 135).

ASK YOUR DOCTOR
SINGLE-GENE DISORDERS

Q **My brother has phenylketonuria, but I do not. Could I transmit the disease to my children?**

A Because phenylketonuria (PKU) is a recessive disorder, both of your parents are carriers. You have a 2 in 3 chance of being a carrier and a 1 in 50 chance of marrying a carrier. The risk for each of your children is roughly 1 in 300, compared with 1 in 10,000 in the general population. However, if at birth your child is found to have PKU, following a restricted (low-protein) diet throughout life can prevent the mental retardation that otherwise results.

Q **My child has sickle cell anemia. My doctor tells me that my children have a 1 in 4 risk. Does this mean that, if I have three more children, none will be affected?**

A No. The risk applies to each pregnancy and is not influenced by the health of any previous children. Each of your children has a 1 in 4 risk of inheriting the disorder; each unaffected child has a 2 in 3 chance of being a healthy carrier.

Q **My son has Duchenne type muscular dystrophy but no one else in my family has the disease. Isn't it inherited?**

A Two thirds of boys inherit the X-linked recessive gene from their mother, who is a carrier. However, the disorder sometimes results from a new gene mutation that occurs during egg or sperm formation; in such cases there is no family history. DNA analysis may help determine whether or not you are a carrier of the gene and at risk of transmitting the disorder to another son.

COULD YOU BE A CARRIER OF A GENETIC DISEASE?

A person is a carrier of a genetic disorder if he or she has the gene for the disorder but is not actually affected by it. In a family affected by an inherited disorder, it may be possible, by studying the family history, to determine which members are carriers. For example, unaffected parents of a child with an autosomal recessive disorder are obvious carriers. Other family members may be carriers as well but nothing in the family history links them to the disorder. For some inherited diseases, it is possible to determine if a person is a carrier through a careful physical examination or laboratory testing. Those people found to be carriers may want to seek genetic counseling before planning a family.

Bill (deceased)   **Louise (78)**

AUTOSOMAL DOMINANT DISORDERS

People who have a dominant gene for a disorder are usually affected. However, there are two ways in which you can be a healthy carrier of a dominant disease gene. In one instance, you have the gene but it has no noticeable effect on you. In the other instance, you are not affected until later in life.

Any person with both a parent and a child with the same dominant genetic disorder is a carrier of the disorder. The children of people with a dominant disorder, as well as the children of carriers, are all possible carriers.

Mary (50)   **Elliot (52)**

CHROMOSOME REARRANGEMENTS

Although the term "carrier" usually refers to single-gene disorders, it is possible for a person to be a carrier of a type of chromosome rearrangement called a translocation. Carriers of this type of chromosome rearrangement are healthy themselves because none of their genetic information is missing. However, their children can inherit the chromosome rearrangement in an unbalanced form that causes illness.

The Smith family
The Smith family is affected by an inherited metabolic disorder called acute intermittent porphyria. Because it is a dominant disorder, it can occur when a person has one defective copy of the gene and one normal copy. The disorder is sometimes not exhibited in a person with the gene, or symptoms may occur only when the person takes certain drugs. Because the disorder has affected both Mary's daughter Jane and her father Bill, Mary can be presumed to be a carrier. Her son Brian and granddaughter Elizabeth are possible carriers. They may benefit from tests to determine whether they carry the disease gene and are at risk of being affected themselves or of transmitting the gene to children.

Brian (28)   **Jane (26)**   **Dean (27)**

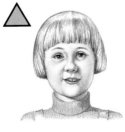

Elizabeth (4)

 Affected

 Unaffected, but a definite carrier

 Neither affected nor a carrier

 Unaffected, but a possible carrier

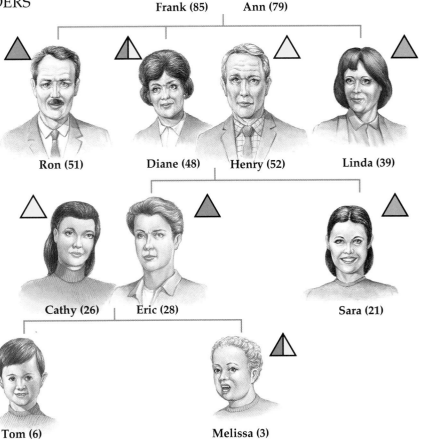

The Taylor family
*Paul Taylor has the autosomal recessive
metabolic disorder phenylketonuria. His
parents and his daughter Cheryl are
therefore carriers; his brother Tim is a
possible carrier. Tim and Cheryl may
benefit from genetic counseling if either
decides to have children.*

AUTOSOMAL RECESSIVE DISORDERS

Carriers of a defective autosomal recessive
gene are themselves healthy. However, if two
carriers of the same defective recessive gene
have children, each child has a 1 in 4 risk of
inheriting the gene from both of them and
therefore of having the disorder. Because an
affected person has two copies of a recessive
gene, his or her parents and children are carriers because they each must have one copy of
the abnormal gene. The brothers and sisters of
anyone affected by an autosomal recessive disorder are possible carriers.

A few autosomal recessive disorders, such
as sickle cell anemia in blacks, are sufficiently
common in certain populations to justify carrier screening in those populations (see
DETECTING CARRIER STATUS on page 107).

X-LINKED RECESSIVE DISORDERS

The carriers of X-linked recessive
disorders are always female. Because
the disease-causing gene may arise
from a new mutation, in order for a
woman to be considered a definite
carrier she must have either two
affected sons, or one affected son and
another affected male relative. The
daughter of a man with an X-linked
recessive disorder is a carrier because she inherits his only X
chromosome, which contains the
defective gene. The female relatives
of a woman with an affected son are
possible carriers.

The Johnson family
*Eric Johnson and his uncle Ron
both have hemophilia, an X-linked
recessive disorder. Therefore,
Eric's grandmother Ann, mother
Diane, and daughter Melissa are
all carriers. His aunt Linda and
sister Sara are possible carriers of
the hemophilia gene; they may
benefit from carrier testing if they
decide to have children.*

GENETICS OF COMMON DISEASES

MANY COMMON CHRONIC DISEASES – including coronary heart disease and the two main types of diabetes (insulin-dependent and non-insulin-dependent) – and some mental disorders have a significant genetic component. Few of these diseases follow a simple, predictable pattern of inheritance. They are seldom caused by a single defective gene, but rather by a combination of many genes and environmental factors.

Some families have a higher-than-average incidence of some common disorders, such as coronary heart disease, asthma, diabetes, and peptic ulcers, and disorders that affect the brain, such as epilepsy and schizophrenia. Although genes play a part in causing these disorders, the relative contribution of genes and environment varies among them.

GENES OR ENVIRONMENT?

Identifying the exact combination of genetic and environmental factors that contributes to many common diseases is difficult, partly because genetically related individuals tend to share the same environment. For some diseases, a

Family risks
The risk of some disorders caused by a combination of genes and environment is greater than average in some families (see graph at right). The blue curve represents the susceptibility of the general population. The red curve represents the susceptibility of people whose genetic makeup makes them more vulnerable to the disease.

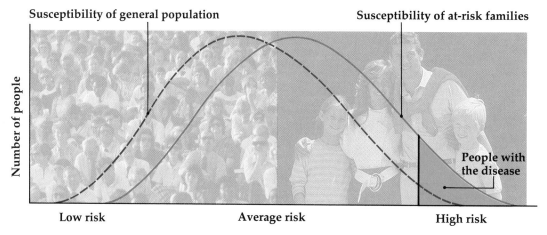

Susceptibility of general population

Number of people

Susceptibility of at-risk families

People with the disease

Low risk   Average risk   High risk

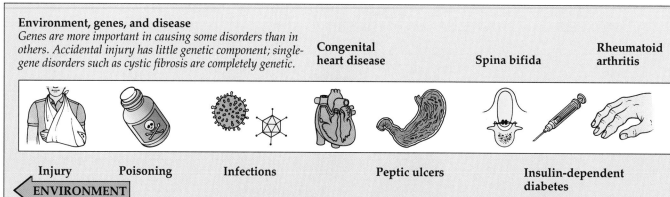

Environment, genes, and disease
Genes are more important in causing some disorders than in others. Accidental injury has little genetic component; single-gene disorders such as cystic fibrosis are completely genetic.

Congenital heart disease   Spina bifida   Rheumatoid arthritis

Injury   Poisoning   Infections   Peptic ulcers   Insulin-dependent diabetes

ENVIRONMENT

definite pattern of inheritance can be established. For example, coronary heart disease is inherited if it results from a gene that causes cholesterol to reach unusually high levels in the blood in a disorder called familial hypercholesterolemia. However, a person who has inherited the gene for hypercholesterolemia can reduce his or her risk of coronary heart disease somewhat by eating a low-fat diet and exercising regularly. For other disorders, the pattern of inheritance is less clear because environmental factors contribute as well. For example, a person in a family that has a history of stomach ulcers may not develop the ulcers unless he or she smokes cigarettes or drinks excessive amounts of alcohol.

HERITABILITY

The heritability of a disease is a rough estimate of the degree to which genes contribute to its development. A heritability of 100 percent indicates that a disease results entirely from the influence of genes. In reality, however, environmental factors are also involved in the development of most common chronic diseases, and their heritability score is much lower than 100 percent.

Estimates of the heritability of common disorders caused by a combination of factors rely heavily on studies of the incidence of the disorder in pairs of identical and nonidentical twins (see TWIN STUDIES on page 22). Estimates are also

based on comparing the disease's frequency in relatives of affected people with the frequency in the general population. Such research suggests, for example, that diseases including asthma, schizophrenia, non-insulin-dependent diabetes, and coronary heart disease have a moderate to high heritability, while peptic ulcers have a low heritability.

Risk factors

Close relatives of a person who has a disorder with a moderate or high heritability are at a higher risk of developing the disorder than a person in the general population. If more than one member of a family is affected by the disorder, the risk to other family members is greater still. The increased risk diminishes as the blood relationship to the affected person becomes more distant. The risk to relatives is sometimes influenced by the age at which a person experienced the first

Predicting risks
For some disorders, genetic counselors can evaluate the risks to a certain family. They do so by studying the incidence of the disorder in the family and comparing it to statistics compiled from the study of many other affected families.

Ankylosing spondylitis   Asthma   Manic depression   Single-gene disorders

Epilepsy   Coronary heart disease   Non-insulin-dependent diabetes   Chromosome abnormalities   GENES

symptoms of the disease and by the severity of the illness in that person. For example, the close relatives of a person who died of a heart attack caused by coronary heart disease are at a much higher risk of dying of a heart attack at an early age if the relative died at age 40 or younger than if he or she died at 65.

IMMUNITY AND DISEASE

Although the precise blend of genetic and environmental factors involved in many common chronic diseases is unknown, the body's immune system is thought to play a significant role in some of them. The presence of proteins called human leukocyte antigens (HLAs) on the surface of cells – which are produced by genes on chromosome 6 – has been linked to some common chronic diseases. For example, specific HLAs are associated with multiple sclerosis, ankylosing spondylitis (see opposite), rheumatoid arthritis, and some types of thyroid disease. Antigens are identification markers that enable the immune system to distinguish the body's tissues from foreign tissues or invading organisms such

DISORDERS ASSOCIATED WITH CHROMOSOME 6

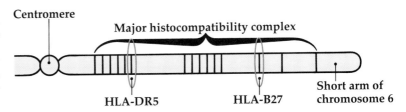

Centromere

Major histocompatibility complex

HLA-DR5

HLA-B27

Short arm of chromosome 6

Hashimoto's disease
Hashimoto's disease develops when the immune system mistakenly attacks the thyroid gland. Symptoms include enlargement of the thyroid gland, muscle weakness, tiredness, and weight loss. The disease is associated with the antigen HLA-DR5, produced by a gene on chromosome 6.

Ankylosing spondylitis
Ankylosing spondylitis is a type of arthritis that causes bones in the spine to fuse together. Studies show 89 percent of affected people have the antigen HLA-B27. A person with the antigen is 100 times more likely to develop the disorder than a person who does not have the antigen.

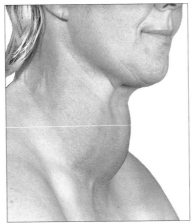

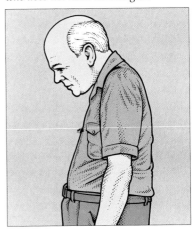

ALLERGIES

Allergic disorders such as asthma and hay fever usually appear in several members of the same family. However, the way in which allergies are inherited varies. The watery eyes and sneezing of hay fever, and the constricted airways that produce the wheezing of asthma, are both caused by an inappropriate immune response. While the susceptibility to becoming allergic is inherited, a specific allergy is not. Often only half the members of an affected family have allergies, which can vary widely in severity.

Skin inflammation
A chronic, itching inflammation of the skin, called atopic eczema (at right), is often found in varying degrees of severity in several members of a family. People with eczema often have other allergic disorders such as asthma.

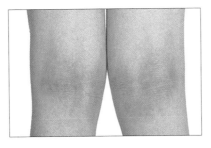

as bacteria and viruses. But, if the immune system fails to recognize the markers, it may mount an attack against its own tissues, causing illness.

Insulin-dependent diabetes

Although the low heritability of insulin-dependent diabetes suggests that environmental factors play a major role in its development, the presence of certain HLAs may contribute as well. The antigens may cause the immune system to mistakenly identify the pancreas as a foreign tissue and attack it, destroying its insulin-producing cells. These HLAs are found in nearly all people with insulin-dependent diabetes. Scientists still do not know how environmental factors trigger the inappropriate immune response in genetically susceptible people.

INHERITED RECEPTOR DEFECTS

On the surface of cells, proteins called receptors are the gatekeepers that recognize and admit other proteins into cells to perform their chemical tasks. In many cases of non-insulin-dependent diabetes, the person has inherited a defect in the receptor for insulin. Insulin helps cells absorb glucose, the sugar that the body uses for fuel. If it is not absorbed by the cell, glucose builds up in the bloodstream, disrupting the body's internal chemistry. Although this type of diabetes is mainly inherited, almost all people who have it are overweight. Many of them can control their blood sugar level with a careful diet and by losing excess weight. Others maintain a safe sugar level with drugs that boost the output of insulin from the pancreas.

Rickets

Rickets, which can affect the skeleton and lead to bone deformities in children, is usually caused by a deficiency of vitamin D in the diet or a lack of exposure to sunlight. But in rare cases, it is inherited. This type of rickets, called vitamin D-resistant rickets, is caused by a defective gene, located on the X sex chromosome, that blocks the formation of the vitamin D receptors in the kidneys. Without the receptors, the kidneys do not convert vitamin D to its active form and it is eliminated from the body.

Vitamin D-resistant rickets
Insufficient absorption of vitamin D makes bones soft and they become curved and stunted. The disorder is inherited in a dominant X-linked pattern. It is more severe in males than females, who have a healthy X chromosome as well as a defective one.

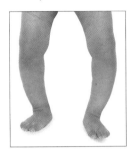

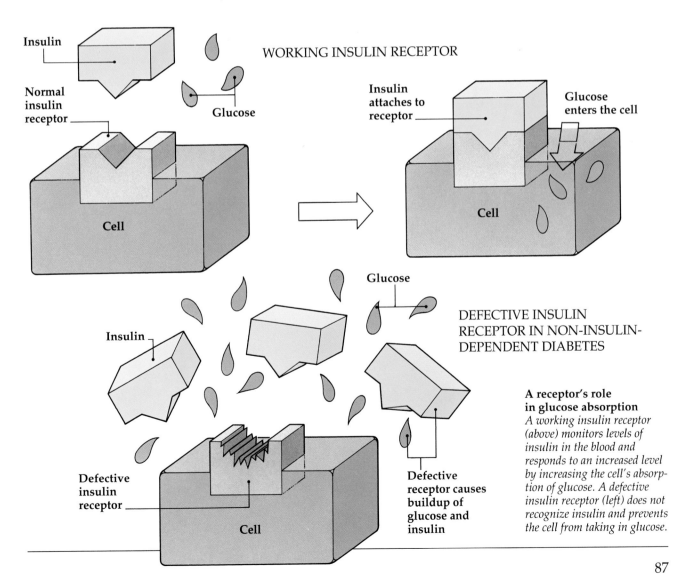

WORKING INSULIN RECEPTOR

Insulin

Normal insulin receptor

Glucose

Cell

Insulin attaches to receptor

Glucose enters the cell

Cell

Glucose

DEFECTIVE INSULIN RECEPTOR IN NON-INSULIN-DEPENDENT DIABETES

Insulin

Defective insulin receptor

Cell

Defective receptor causes buildup of glucose and insulin

A receptor's role in glucose absorption
A working insulin receptor (above) monitors levels of insulin in the blood and responds to an increased level by increasing the cell's absorption of glucose. A defective insulin receptor (left) does not recognize insulin and prevents the cell from taking in glucose.

ARE YOU AT RISK OF DIABETES?

Diabetes has both genetic and environmental influences. There are two main types: insulin-dependent and non-insulin-dependent. In insulin-dependent diabetes, the cells in the pancreas cannot produce the hormone insulin. Insulin enables cells to absorb an essential fuel: sugar in the form of glucose. When the cells cannot absorb glucose, sugar accumulates in the blood. In non-insulin-dependent diabetes, the high blood sugar level is not caused by a deficiency of insulin, but rather by a malfunction in how the body responds to insulin. Being overweight is an important triggering factor for this type of diabetes.

SYMPTOMS OF DIABETES

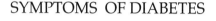

Blurred vision
If diabetes is not controlled, the high blood sugar level can damage the light-sensitive area at the back of the eye, resulting in blurred vision or even blindness.

Nerve damage
Many people with diabetes eventually have tingling, weakness, and loss of sensation in the arms and legs from nerve damage.

Heart disease
Coronary heart disease is much more common in people with diabetes than in the general population because diabetes causes damage to blood vessels.

Frequent urination
Excess sugar is excreted in the urine. Frequent urination and tremendous thirst are often the first signs of diabetes.

Cystitis
The high level of sugar in the blood interferes with the body's ability to fight infections such as cystitis (inflammation of the bladder) and skin infections.

Muscle weakness and fatigue
A person with diabetes feels tired and weak because the body's cells are deprived of their source of energy – glucose.

Pain in the legs
In people who already have atherosclerosis (a buildup of fat in the lining of arteries), diabetes causes further blockage. Reduced blood flow into leg muscles can cause pain when walking.

Insulin-dependent diabetes
Insulin-dependent diabetes can develop at any age but it usually develops between the ages of 10 and 16. The insulin-secreting cells in the pancreas stop producing the hormone, causing sugar (glucose) to build up in the blood. In the US, four out of every 1,000 people have insulin-dependent diabetes. All people with untreated insulin-dependent diabetes have symptoms.

Non-insulin-dependent diabetes
Ninety percent of people with diabetes have non-insulin-dependent diabetes. This form of diabetes occurs mostly in people 40 or older whose cells have become resistant to the effects of insulin. Obesity, drugs, and some illnesses can trigger the illness. Half of the 13 million Americans who have this type of diabetes do not know they have it. Five out of every 100 people in the US develop non-insulin-dependent diabetes.

TREATING DIABETES

Although diabetes cannot be cured, its symptoms can be alleviated by keeping the level of glucose in the blood within the normal range.

Diet and exercise
People with diabetes can help control their blood sugar level by monitoring the amount of carbohydrates and simple sugars (such as those found in candy and desserts) they eat. For people with non-insulin-dependent diabetes, a low-calorie, low-fat diet; loss of excess weight; and exercise may be enough to maintain a safe blood sugar level.

Insulin injections
To keep their blood sugar level down, people with insulin-dependent diabetes must give themselves daily injections of insulin.

GENETIC INFLUENCES IN DIABETES

Genetic factors play a part in the development of both types of diabetes. However, genes are a more significant component in non-insulin-dependent diabetes. Your risk of diabetes depends on which of your relatives have the disease and which type they have.

RISK OF DIABETES	NO FAMILY HISTORY	ONE PARENT AFFECTED	BOTH PARENTS AFFECTED	SIBLING AFFECTED	IDENTICAL TWIN AFFECTED
Insulin-dependent diabetes *A person with one parent or a sibling with insulin-dependent diabetes has less than a 10 percent risk of developing it. Still, that risk is up to 25 times higher than the risk for a person with no family history of insulin-dependent diabetes.*	0.4%	1% if mother affected 6% if father affected	20%	3 to 10%	50%
Non-insulin-dependent diabetes *A person with a parent or sibling with non-insulin-dependent diabetes has a moderate risk of developing it. Nearly 40 percent of siblings of people with non-insulin-dependent diabetes develop it by the age of 80.*	5%	10% if mother affected 10% if father affected	40%	10 to 40%	97%

89

CORONARY HEART DISEASE

You can inherit a susceptibility to heart disease, but environmental factors – such as diet, exercise, and cigarette smoking – can modify that risk. As a person ages, a mixture of fat and cholesterol tends to build up in the inner walls of the arteries. If these fatty deposits eventually block one of the coronary arteries, the person is at risk of having a heart attack.

Clogged arteries
The buildup of deposits of fat and cholesterol in the coronary arteries can eventually prevent blood flow to the heart and lead to a heart attack.

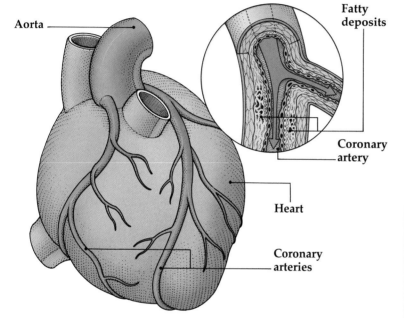

Aorta

Fatty deposits

Coronary artery

Heart

Coronary arteries

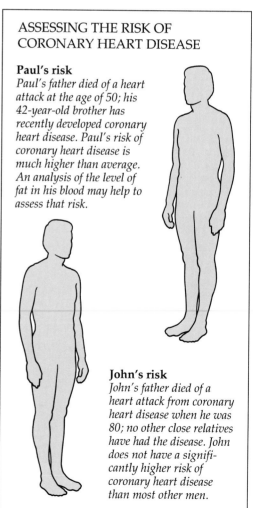

terolemia is caused by a single abnormal, dominant gene that is carried by about one person in 500. People who inherit two copies of the harmful gene may be affected so severely that they have a heart attack in early childhood.

Genetic risk factors

Genetic factors appear to have a stronger influence on a man's susceptibility to heart disease than on a woman's. But any person's risk of coronary heart disease is increased if he or she has a close relative with coronary heart disease or one who has died of it. That risk increases if more than one relative has heart disease. If any of your close relatives has died of heart disease at an early age, or if any relative with the disease is female, you are at still greater risk.

ASSESSING THE RISK OF CORONARY HEART DISEASE

Paul's risk
Paul's father died of a heart attack at the age of 50; his 42-year-old brother has recently developed coronary heart disease. Paul's risk of coronary heart disease is much higher than average. An analysis of the level of fat in his blood may help to assess that risk.

John's risk
John's father died of a heart attack from coronary heart disease when he was 80; no other close relatives have had the disease. John does not have a significantly higher risk of coronary heart disease than most other men.

Severe familial hypercholesterolemia
Yellowish accumulations of fat under the skin are a common feature of severe familial hypercholesterolemia.

Fat and heart disease

You can inherit disorders that impede your body's ability to use and control the level of cholesterol in your blood, making you susceptible to heart disease. Eating foods high in saturated fats, which form cholesterol in the blood, may lead to a buildup of fatty deposits in your heart's arteries. Even the structure of your coronary arteries, which is determined by genes, can influence the formation and buildup of cholesterol. Some people have an inherited disorder called familial hypercholesterolemia that prevents their body from adequately regulating the level of fats and cholesterol in their blood, resulting in a high blood cholesterol level. Familial hypercholes-

MULTIPLE SCLEROSIS

Multiple sclerosis is a chronic and progressive disorder of the central nervous system. Genes play a part in multiple sclerosis, but their exact role is unknown. Symptoms can range from numbness and tingling to paralysis. The illness usually occurs in increasingly severe episodes of varying duration. Environmental factors, such as viruses, are thought to be important in triggering multiple sclerosis.

Nerve damage
Multiple sclerosis gradually destroys myelin, the protective covering of nerve cell axons (signal-conducting fibers). Nerve cells cannot function properly without myelin.

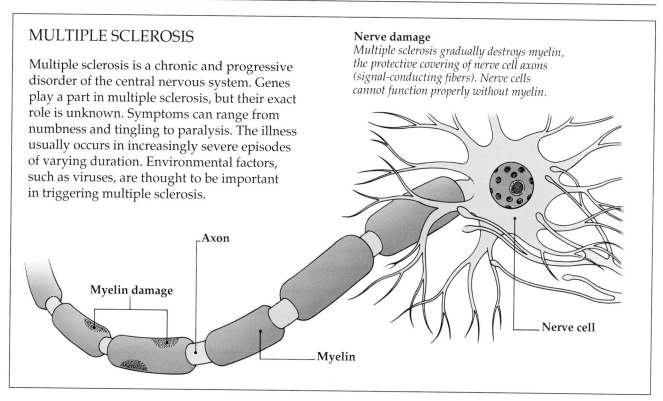

Axon

Myelin damage

Nerve cell

Myelin

RHEUMATOID ARTHRITIS

Rheumatoid arthritis is a disease that causes inflammation of the joints, particularly in the hands, which may eventually lead to severe disability. When larger joints, such as the hip and knee joints, are affected, they can often be successfully replaced with artificial ones. Rheumatoid arthritis is more common in women than in men and usually develops in early adulthood or in middle age.

More than two thirds of people with rheumatoid arthritis have a specific HLA, which is a protein identification marker, on the surface of their cells (see IMMUNITY AND DISEASE on page 86). This particular marker is six times more common in families with more than one affected member. If you have a parent or sibling with rheumatoid arthritis, your risk of developing the disease is 1 in 20; if the relative is male, your risk is even higher.

Arthritic hand
Joints affected by rheumatoid arthritis become swollen, painful, and stiff. In severe cases, the finger joints may be completely destroyed. Taking anti-inflammatory drugs usually helps relieve the symptoms.

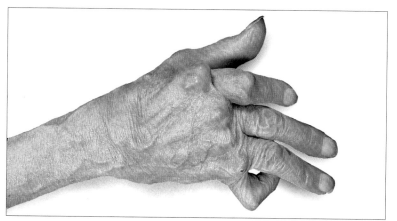

SCHIZOPHRENIA

Schizophrenia is a mental disorder characterized by hallucinations and delusions, inappropriate emotional reactions, and an inability to concentrate. Schizophrenia is thought to have a high heritability. If one member of a pair of identical twins has schizophrenia, the other twin has up to a 50 percent risk of also developing it. Studies of identical twins raised apart have shown similar results. For a person with an affected parent, brother, or sister, the risk of schizophrenia is about 10 to 15 percent. If both parents are affected, the risk increases to 40 percent. The incidence of schizophrenia in the general population is 1 percent.

ALZHEIMER'S DISEASE

Dementia – progressive deterioration of a person's mental abilities – can result from a variety of illnesses. An estimated 10 to 15 percent of people older than 65 have Alzheimer's disease, the most common form of dementia. The exact cause of Alzheimer's disease is not fully understood. One form is thought to be caused by a single abnormal gene. This more severe, inherited form of Alzheimer's disease, which develops at an earlier age (about 50) and progresses more quickly, is linked to a defect on chromosome 21. In fact, people who survive beyond age 30 with Down's syndrome, which results from the presence of an extra chromosome 21, develop a form of dementia indistinguishable from Alzheimer's disease. Each child of a person with the inherited, early-onset form of Alzheimer's disease has a 50 percent chance of also developing it.

Most people with Alzheimer's disease have a form that is much less strongly linked to genes. If only a single case of Alzheimer's disease has occurred in a family, the risk to sisters, brothers, and children is only slightly higher than the risk in the general population.

PARKINSON'S DISEASE

Parkinson's disease is a brain disorder that affects many older Americans. Features of the disorder include a rigid posture, slow movements, a tremor, and a shuffling, unsteady walk. A third of

Loss of brain tissue
Alzheimer's disease causes brain cells in some areas of the brain to degenerate. The resulting loss of nerve tissue causes the brain to slowly shrink, leading to dementia.

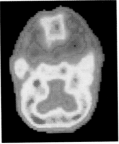

Normal brain

Brain with advanced Alzheimer's disease

Diagnosing epilepsy
To diagnose epilepsy, doctors study recordings, called electroencephalograms (EEGs), of the electrical activity of the person's brain. The pattern of electrical activity changes significantly during a seizure (see below).

Manic depression
Manic depression is a mood disorder that appears to have a strong genetic influence. People with manic depression have alternating periods of severe depression and extreme, often irrational, excitement.

people with Parkinson's disease eventually lose some mental ability. Although it is believed that some people inherit a susceptibility to Parkinson's, it is not known what genetic or environmental influences trigger the illness. For an isolated case, the risk to close relatives is only slightly greater than the risk of the general population. However, a person with two relatives with Parkinson's disease has a 30 percent chance of developing it by age 75. Although the disease cannot be cured, medication can help to minimize the symptoms.

EPILEPSY

Epilepsy – characterized by seizures (periods of increased electrical activity in the brain) – is a disorder that affects nearly one in 200 people. Epilepsy can result from factors such as brain injury or infection, but usually the cause cannot be determined. In a relatively small percentage of cases, epilepsy is inherited as part of a genetic disorder such as tuberous sclerosis (see page 100).

One form of epilepsy, called petit mal epilepsy, has a strong genetic influence. Petit mal seizures consist of short periods of absence of attention during which a person stares vacantly into space. Many of these children outgrow the disorder after adolescence. Half of the brothers and sisters of a person with petit mal epilepsy have similar electrical distur-

Normal EEG

EEG in petit mal seizure

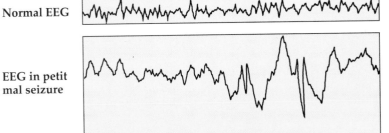

bances in their brains but only 6 percent actually have seizures.

In the more common form, grand mal epilepsy, the seizures are more severe. A person may lose consciousness during a seizure and may even momentarily stop breathing and turn blue. This type of epilepsy is less influenced by genes and is more likely to have been caused by an injury to the brain. Anticonvulsant drugs can prevent or decrease the frequency of seizures in most people with epilepsy.

Hereditary factor in grand mal epilepsy
If one parent has grand mal epilepsy (shown here in green), each child's risk is about 5 percent; if both parents are affected, the risk is 15 percent. Siblings of a child with grand mal epilepsy have a 5 percent risk.

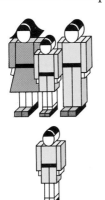 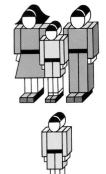 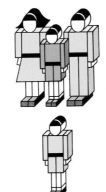

| 5 percent risk | 15 percent risk | 5 percent risk |

ALCOHOLISM AND GENES

Studies indicate that the tendency to become addicted to alcohol is influenced by genes. People vary in the way their body deals with alcohol. For example, the speed with which the liver removes alcohol from the body may influence a person's risk of addiction. Still, it is hard to separate genetic factors from the many environmental influences that might make a person become addicted. Genetically susceptible people can prevent addiction by not drinking alcohol.

Environmental influences
If members of a person's family or social group drink regularly, he or she is also likely to drink. In genetically susceptible people, the risk of alcohol addiction increases with the quantity of alcohol routinely consumed.

ASK YOUR DOCTOR
GENETIC RISKS

Q My brother is severely mentally retarded. Is there a chance that my husband and I could have a mentally retarded child?

A Your risk of having a mentally retarded child depends on the cause of your brother's condition. If his retardation results from the inherited disorder fragile X syndrome, or some other X-linked disorder, each of your sons has as much as a 1 in 4 risk of being retarded. Your daughters will be unaffected. However, if your brother's retardation is not genetic, your children's risks may be no greater than those of any other children. Ask your doctor about genetic counseling; a genetic counselor can help you evaluate your risks.

Q My wife has insulin-dependent diabetes. Are our children likely to be diabetic too?

A Although insulin-dependent diabetes is the less inherited form of diabetes, each of your children has a 1 in 100 chance of developing it. That risk is about two to three times greater than the risk in the general population.

Q My sister, who is 35 years old, has multiple sclerosis. I am 25 and I am worried that I will also get it. What are my chances?

A Right now, your risk of developing the chronic central nervous system disorder known as MS is about 1 in 20 but, by age 42, your risk decreases to less than 1 in 100. Multiple sclerosis is caused by a number of factors, few of which are fully understood. It is thought to result from the interaction of an environmental agent, such as a virus, with a genetic susceptibility.

BIRTH DEFECTS

ABNORMALITIES THAT ARE PRESENT at birth are called birth defects or congenital abnormalities, even though their effects are sometimes not apparent until later in life. Birth defects may result from either genetic or environmental factors, or a combination of the two. A simple explanation for the cause of an abnormality often cannot be found. Each year in the US, nearly 250,000 babies are born with birth defects, the major killer of children under age 1.

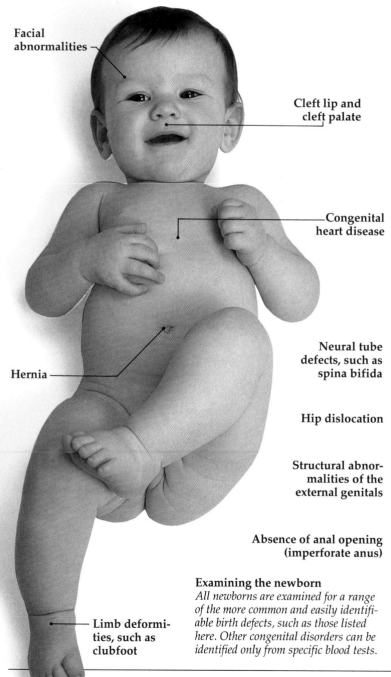

Facial abnormalities

Cleft lip and cleft palate

Congenital heart disease

Hernia

Neural tube defects, such as spina bifida

Hip dislocation

Structural abnormalities of the external genitals

Absence of anal opening (imperforate anus)

Limb deformities, such as clubfoot

Examining the newborn
All newborns are examined for a range of the more common and easily identifiable birth defects, such as those listed here. Other congenital disorders can be identified only from specific blood tests.

A fetus undergoes tremendous growth and development during the 9 months of pregnancy. Any interruption in the development process during that time can result in a birth defect.

TYPES OF BIRTH DEFECTS

A congenital abnormality or birth defect is not the same as a genetic disorder. Some birth defects, such as polydactyly (extra fingers or toes), are associated with single-gene disorders, and some result when chromosomes are rearranged during cell division when eggs or sperm are formed (see page 60). However, some birth defects result from unknown causes, and these abnormalities rarely occur twice in the same family. Other birth defects have no genetic component and are caused exclusively by environmental factors. For example, the baby of a pregnant woman with syphilis may be born with congenital syphilis, acquired during fetal development.

Multiple factors
Many of the more frequent birth defects, such as spina bifida, are thought to result from the interaction of several genes and specific environmental influences. Even though these abnormalities are not entirely inherited, they occur more often in an affected child's brothers and sisters than in the general population. This finding suggests that genetic predisposition plays a role in the occurrence of these birth defects in some families.

CAUSES OF BIRTH DEFECTS

Some types of birth defects occur as the result of a number of factors that interfere to some degree with the normal development of a fetus.

Disruptions in development

Disruptions cause destruction to a part of a fetus that was previously developing normally. For example, a tear in the membrane of the amniotic sac surrounding the fetus may result in a membrane strand floating free in the amniotic fluid and wrapping around a fetus's arm, toes, or fingers. If the blood supply is cut off, these body parts may never develop. These children are genetically normal and the risk of such a defect occurring in a subsequent pregnancy is very small.

Deformations

Deformations can result when the movement of a fetus inside the uterus is restricted, causing the fetus to remain in an abnormal position for a long time. Such restriction may be caused, for example, by an irregularly shaped uterus or a multiple pregnancy in which more than one fetus is competing for space.

The outlook for these babies is usually good because the abnormality, such as feet that turn in or out, can be relatively easy to correct or may even correct itself. The risk of having another child with a deformation is low unless the abnormality was caused by an abnormally shaped uterus.

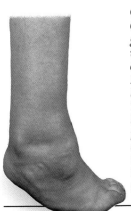

Clubfoot
Clubfoot (left) is a congenital deformation in which the foot is twisted out of shape or position. Five in 1,000 babies are born with the defect. Treatment begins at birth and includes physical therapy or a cast to correct the distortion of soft tissues and bones. Surgery is rarely performed.

Malformations

Malformations are structural defects that occur during organ and tissue formation in early fetal development. They can be single or multiple and relatively minor or life-threatening. Single malformations, which are relatively common, include cleft lip and palate, congenital heart defects, and neural tube defects (such as spina bifida). The risk that a single malformation will occur again in a family is greater than the risk in the general population. During pregnancy, ultrasound scans can detect many kinds of single malformations.

NORMAL ANATOMY   SPINA BIFIDA

Nerve root — **Spinal cord** — **Skin** — Cerebrospinal fluid — **Vertebra** — **Missing spinal bone** — **Vertebra**

Spina bifida
In the most serious form of spina bifida (below), a baby is born with an exposed swelling over the spine (consisting of nerve tissue of the spinal cord) that may or may not be contained in a membrane. This child is likely to be severely handicapped.

RISKS FOR SOME BIRTH DEFECTS

DISORDER	INCIDENCE PER 100 BIRTHS	RISK (PERCENT) OF OCCURRENCE IF ONE PARENT IS AFFECTED	RISK (PERCENT) OF RECURRENCE IF ONE CHILD IS AFFECTED
Cleft lip and palate	0.1	4	4
Cleft palate alone	0.04	7	2
Clubfoot	0.1	3	3
Congenital dislocation of the hip	0.07	12	6
Congenital heart defects	0.5	1 to 4	1 to 4
Hydrocephalus (excess amount of fluid within the skull)	0.05	–	3
Neural tube defects	0.1 to 0.2	2 to 3	2 to 3

CASE HISTORY
RISK OF SPINA BIFIDA

SANDRA AND HER TWO **sisters are all healthy, but their brother died at the age of 1 month from complications of severe spina bifida. Sandra and her husband have decided to start a family but Sandra is worried that she will have a child with spina bifida. The couple decide to ask their doctor for advice.**

PERSONAL DETAILS
Name Sandra Williams
Age 25
Occupation Journalist
Family Sandra's brother was the only family member known to have had spina bifida.

MEDICAL BACKGROUND
Sandra was 3 when her brother was born with spina bifida. Sandra was then examined by a doctor and found to be free of any spinal cord abnormalities. She and her husband are both healthy.

THE CONSULTATION
Sandra tells the doctor of her fears. The doctor explains that spina bifida is one of a range of related congenital (present at birth) disorders called neural tube defects, which result when an embryo's nerve tissue fails to fold together to form the tube that becomes the brain and spinal cord. Although genes play a role, neural tube defects are not completely inherited. Environmental influences, including a pregnant woman's diet, may also play a part in their de-velopment. Because of her family history, each of Sandra's children would have a higher-than-average risk of a neural tube defect. That risk is about 1 percent. The doctor explains that a deficiency of folic acid may be associated with neural tube defects and suggests that Sandra take folic acid tablets daily while she is trying to get pregnant and throughout the first 3 months of her pregnancy. It is during the first 3 months of pregnancy that the neural tube develops. If Sandra and her husband do have a child with a neural tube defect, each of their subsequent children will have about a 2 to 3 percent risk of also having one. The doctor tells the couple that tests are available to detect neural tube defects during pregnancy. These tests include ultrasound scanning, blood tests, and amniocentesis.

FURTHER INVESTIGATION
Sandra and her husband decide to go ahead with their plans to start a family and Sandra begins taking folic acid supplements. Three months later she becomes pregnant and continues eating a healthy diet and taking the folic acid supplements. A blood test (see below) at 16 weeks indicates that the fetus is unlikely to have a neural tube defect. Those results are confirmed by an ultrasound scan a few days later.

THE OUTCOME
Sandra has a healthy boy. Reassured that he is normal, Sandra and her husband begin thinking about having more children.

Blood test
At 16 weeks of pregnancy, Sandra's blood is tested for the level of alpha-fetoprotein, a substance made by the fetus's liver. A high level of the substance indicates an increased risk of a neural tube defect. The level of alpha-fetoprotein in Sandra's blood indicates that the risk of spina bifida in the fetus is low.

Syndromes

Some malformations occur together more often than would be expected by chance. These multiple malformations, which are called syndromes, are not as common as single malformations and may have many causes, both environmental and genetic, or may have no known cause. They are among the most serious congenital abnormalities.

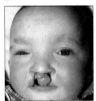

Cleft lip and palate
Cleft lip is a vertical split in the upper lip that may extend up to the nose (above left) and can occur with cleft palate (a gap in the roof of the mouth). The child in the photograph below left has had reconstructive surgery, which can almost completely correct these abnormalities.

PYLORIC STENOSIS

In a baby born with pyloric stenosis the circular muscle at the outlet of the stomach is thickened, obstructing the passage of food into the intestine. The disorder, which mostly affects male babies, causes severe vomiting about 3 to 4 weeks after birth. The estimated risks to family members are illustrated below.

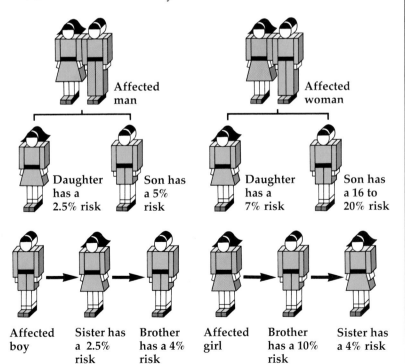

Affected man

Daughter has a 2.5% risk
Son has a 5% risk

Affected woman

Daughter has a 7% risk
Son has a 16 to 20% risk

Affected boy
Sister has a 2.5% risk
Brother has a 4% risk

Affected girl
Brother has a 10% risk
Sister has a 4% risk

ASK YOUR DOCTOR
CONGENITAL ABNORMALITIES

Q My son was born with a heart defect. Did this happen because I was so shocked by my father's death during my pregnancy?

A No. There is no evidence that emotional trauma during pregnancy can cause a congenital abnormality in a child.

Q My first child was born with a diaphragmatic hernia and lived for only 48 hours. My doctor says that the chance of this happening again is small. Are there any tests I can have during pregnancy?

A Yes. Although your risk of having another affected child is low, an ultrasound scan during pregnancy can detect diaphragmatic hernia (in which abdominal organs protrude through the diaphragm into the chest). In addition to eliminating the possibility of this problem, a normal scan will reassure you that the fetus is unlikely to have any other major structural defects.

Q I was born with a cleft lip and palate. Will my children also have these abnormalities?

A Not necessarily. Your children's risk depends on a number of factors. If you have no other congenital abnormalities, and none of your relatives has the defect, the risk to each of your children is about 1 in 25. The risk is higher if other relatives are affected. Cleft lip and palate may be caused by drugs taken during pregnancy, but, in these cases, other abnormalities are usually present as well. If drugs caused your condition, your child's risk is almost the same as that of the general population.

GENES AND CANCER

CHANGES OR MUTATIONS in the DNA of cells can lead to cancer. Some of these mutations are inherited directly, but most are caused by by-products of the body's chemistry or environmental toxins. It is now thought that, out of the approximately 100,000 genes in each of your cells, more than 20 specific genes can lead, as a result of mutation, to the development of cancer.

Cancer is the unregulated division of cells. Although most types of cancer are not actually hereditary, a person can inherit a susceptibility to some cancers.

CANCER-CAUSING GENES

Two types of genes are thought to be involved in the development of cancer: proto-oncogenes and antioncogenes.

Proto-oncogenes promote the normal growth and division of cells. Most of the time these genes are inactive, or switched off. But a mutation in a proto-oncogene can permanently switch it on, turning it into an oncogene. Oncogenes, which operate like gas pedals that are stuck, lead to the excessive cell division that is characteristic of cancer. Antioncogenes normally serve as brakes to keep cell division in check. But when a mutation inactivates an antioncogene, it no longer performs its job of preventing runaway cell division and cancer results.

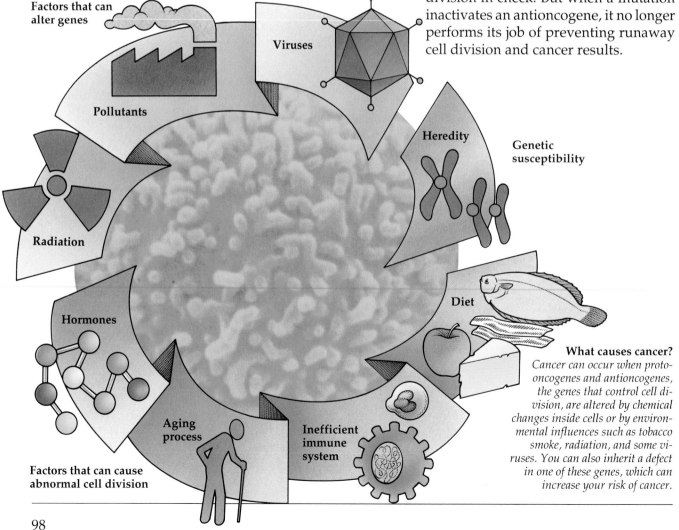

Factors that can alter genes

Viruses

Pollutants

Heredity

Genetic susceptibility

Radiation

Diet

Hormones

Aging process

Inefficient immune system

Factors that can cause abnormal cell division

What causes cancer?
Cancer can occur when proto-oncogenes and antioncogenes, the genes that control cell division, are altered by chemical changes inside cells or by environmental influences such as tobacco smoke, radiation, and some viruses. You can also inherit a defect in one of these genes, which can increase your risk of cancer.

CANCER AND HEREDITY

If you have a relative with cancer or one who died of cancer, you are not necessarily at increased risk of developing it yourself. Some types of cancer, such as cancer of the cervix, lung, or bladder, are rarely inherited. Nevertheless, some families appear to have a susceptibility to a certain cancer. Between 4 and 10 percent of all cases of cancer of the breast, colon, and ovary are thought to have a genetic component. At-risk families cannot be identified until a particular cancer has occurred more than once in the family and in a person of younger-than-average age for that cancer.

Genetic susceptibility to cancer

Usually, cancer develops only after a proto-oncogene or an antioncogene has undergone one or more changes. A person who inherits an already altered gene has an increased risk of cancer. When these altered genes are exposed to certain internal or external influences, such as hormones inside the body or radiation from sunlight, cancer may develop.

In some cancers, inherited mutations make DNA unable to repair itself after it has been damaged. Efficient repair of genetic damage protects the body from cancer. Because people vary in their ability to repair DNA, their risk of developing some cancers depends on how efficient their repair mechanisms are.

SINGLE-GENE DISORDERS AND CANCER

A few single-gene disorders – including neurofibromatosis, familial polyposis, and tuberous sclerosis – make people susceptible to some types of cancer. These disorders usually follow a pattern of autosomal dominant inheritance, which

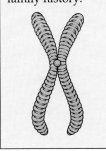

BREAST CANCER
Geneticists have identified a gene on chromosome 17 that causes an inherited form of breast cancer in women under 40. Women with a family history of early breast cancer have a more than 30 percent chance of developing it by age 40, compared with a less than 1 percent risk for women with no family history.

Ovarian cancer
As shown in the family inheritance pattern below, ovarian cancer sometimes affects several women in a family and probably results from an abnormal dominant gene. In these families, the daughters of an affected woman have up to a 50 percent chance of also developing ovarian cancer. Researchers are working to identify the cancer-causing gene, which promises to improve the ability to predict a woman's risk.

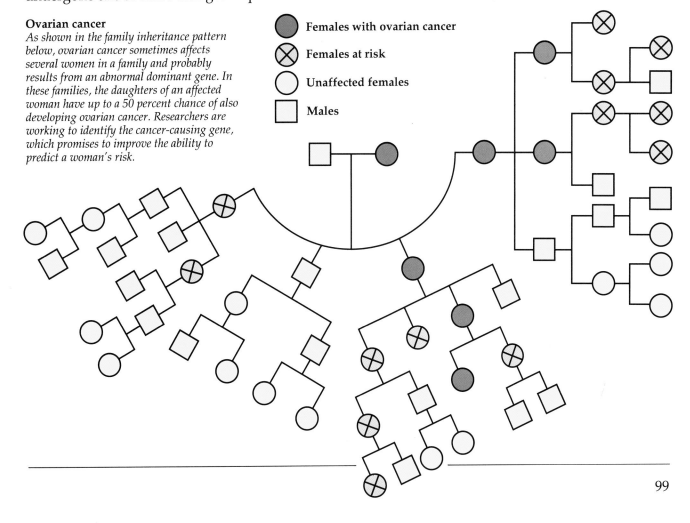

Females with ovarian cancer

Females at risk

Unaffected females

Males

means that only one gene is necessary to cause the disorder. Each child of a person with the abnormal gene has a 50-50 chance of inheriting it, and, as a result, of being susceptible to the cancer.

Neurofibromatosis

Approximately one in every 3,000 people has an abnormality in the gene that regulates the division of nerve cells. The disorder that can result, called neurofibromatosis, produces patches of brownish pigment on the skin and noncancerous tumors that appear as bumps on the skin. One of these benign tumors may develop into cancer and spread. Occasionally, tumors or cancerous growths form on a person's brain and spinal cord. The effects of neurofibromatosis vary from hardly noticeable to severe; nine out of 10 affected people do not develop cancer.

Familial polyposis

About one person in 10,000 has familial polyposis, an inherited disorder characterized by multiple abnormal growths called polyps inside the intestines (mainly the large intestine). This disorder is caused by an altered gene that controls the division of cells lining the intestines. People with familial polyposis tend to develop the polyps during their late teens and early 20s. By age 30, they may have thousands of polyps. Because these cells have already undergone the genetic changes that can lead to cancer, the risk that the polyps will become malignant is high. Without preventive treatment, such as surgical removal of the polyps or even removal of the colon itself, a person with the disorder usually develops colon cancer by the age of 40.

Tuberous sclerosis
Tuberous sclerosis is a genetic disorder with a dominant pattern of inheritance. The disorder can cause abnormal growths in tissues and organs, including the skin, brain, kidneys, and heart. The severity of symptoms varies among affected people; cancerous tumors occur in a small percentage of people.

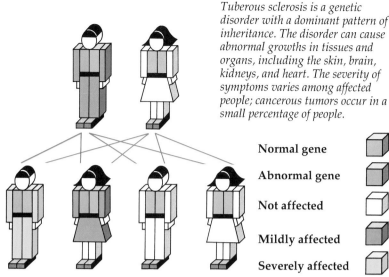

Normal gene

Abnormal gene

Not affected

Mildly affected

Severely affected

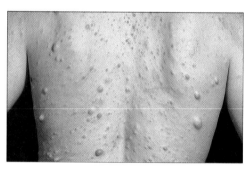

Signs of neurofibromatosis
A person with neurofibromatosis has patches of pigmented skin and may develop numerous soft, fibrous swellings that grow on the nerves in the skin.

Preventive treatment of polyposis
In people with familial polyposis, doctors monitor the development of polyps by performing a colonoscopy, an examination of the colon using a flexible viewing instrument.

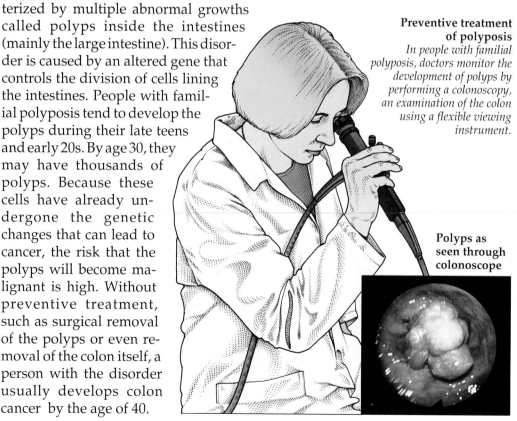

Polyps as seen through colonoscope

CHROMOSOMES AND CANCER

Chromosome defects are characteristic of some forms of cancer. For example, about 90 percent of people who develop chronic myeloid leukemia carry a chromosome rearrangement known as the Philadelphia chromosome. The abnormal chromosome consists of chromosome 9 with a small piece of chromosome 22 attached. The defect tends to switch on a cancer-causing gene (an oncogene), which results in this type of leukemia.

Retinoblastoma in one eye

Retinoblastoma appears initially (often in childhood) as a whiteness in the eye. About 90 percent of small tumors that occur in only one eye can be treated effectively with either surgery or radiation therapy.

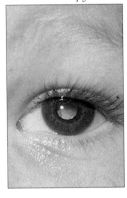

Retinoblastoma

Cancers that appear early in childhood are more likely to have been inherited than cancers that appear later in life. About half of all cases of retinoblastoma, which is cancer of the retina (the light-sensitive layer at the back of the eye), are inherited as the result of an altered anti-oncogene (one of the "brakes" that halt cell division) found on chromosome 13. In about one fourth of cases, children are born with a tumor in each eye. Brothers and sisters of a child with retinoblastoma in both eyes have a 40 percent risk of having it if one of their parents is also affected. Occasionally, a child with no family history develops only one eye tumor. The relatives of that child, including his or her future children, have a low risk of developing retinoblastoma.

DETECTING INHERITED TUMORS

Because the symptoms of neurofi-bromatosis and tuberous sclerosis are apparent on the surface of the body, people with these disorders are usually identified before age 10. Close family members who may also be at risk can then be examined for the disorder.

Carriers of some inherited diseases, such as familial polyposis, can be identified using DNA analysis. The DNA in blood samples taken from family members is examined for the disease-causing gene (see GENETIC ANALYSIS on page 108). If the gene is not found in a blood sample, that person probably needs no further examination and his or her children are at little risk of developing polyposis.

CANCER MONITORING FOR AT-RISK FAMILIES

Although a strong family history may implicate heredity as the cause of a cancer, this is not always the case. Even if the tendency to develop a cancer is inherited, it is usually not possible to identify the abnormal gene and, therefore, the members of the family who may be at risk. All members of an at-risk family should be examined regularly for the first signs of cancer. The earlier a cancer is detected, the more effective the treatment is likely to be. If you have a family history of a cancer, your doctor may recommend that you have regular tests to detect cancer at an early, treatable stage.

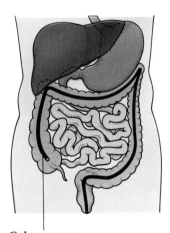

Colonoscope

X-ray machine

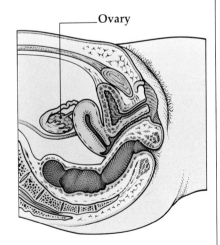

Ovary

Examining the colon
If your family has a risk of colon or rectal cancer, your doctor may recommend that you have a colonoscopy (an examination of the entire colon to detect polyps) every year.

Examining the breasts
Women of any age with a family history of breast cancer should agree on a plan with their doctor to have frequent breast examinations and mammograms. They should also do a monthly self-examination of their breasts.

Examining the ovaries
To check for ovarian cancer, doctors use clinical examination and ultrasound (which is done to obtain an image of the ovaries). An ultrasound scan can detect ovarian cancer at an early stage.

CHAPTER FOUR

GENETIC COUNSELING, DIAGNOSIS, AND TREATMENT

AN ESTIMATED 7 percent of children are born with a birth defect, mental retardation, or a genetic disorder. These defects are the major killer of children in the first year of life. Forty percent of defects present at birth are known to result from genes alone or from a combination of genetic and environmental factors. Although the causes of the remaining 60 percent have not yet been determined, many of these abnormalities may be influenced, at least in part, by genes. Some genetic disorders – such as Huntington's chorea and some forms of blindness and deafness – do not show up until later in life; this type of disorder affects one in 80 adults.

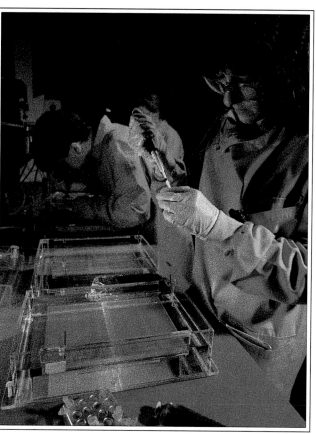

New knowledge about the genes that cause disease and the ways in which birth defects and genetic disorders occur has helped doctors make more accurate diagnoses. They can perform specific tests that allow them to analyze a person's genetic makeup, inform people about how they can avoid transmitting a genetic disorder to offspring, and provide treatment for many disorders.

When a couple has a child with a birth defect or an inherited disorder, the first step is to get a diagnosis. Many disorders have similar symptoms but very different causes. In many cases, it is possible to obtain a laboratory diagnosis. If a child's defect is shown to have a genetic cause, the parents may be referred to a genetic counselor. The counselor can provide them with the known facts about the disorder and offer them more tests, if necessary. The counselor also evaluates a couple's risk of having another child with the same disorder.

Genetic counseling also benefits people who are themselves affected by a genetic disorder or have a relative who is. They may want to know their risk of passing an abnormal gene on to children. Members of some ethnic groups are at increased risk for particular genetic disorders, such as blacks for sickle cell anemia and Ashkenazi Jews for Tay-Sachs disease. They may decide to be tested to find out whether they carry the disease-causing gene. Couples found to be at increased risk of having a child with a serious genetic disorder can choose from a variety of family planning options. Prenatal tests during pregnancy can detect a number of abnormalities in a fetus. If a test diagnoses a severe, untreatable disorder, a couple may choose to terminate the pregnancy. Prenatal detection can also help prepare parents psychologically for the birth of a child with a genetic defect and give hospital staff time to make preparations to care for an affected child immediately after birth.

EVALUATING YOUR RISK

IF YOU ARE AT INCREASED risk of having a child with a genetic disorder, tests can determine whether or not you are a carrier of the disease-causing gene. The test results are used by genetic counselors to evaluate the chances of your transmitting a defective gene to your children. If you are found to be at risk of having a child with a genetic disorder, you might choose an alternative route to parenthood, such as artificial insemination or adoption.

WHO CAN BENEFIT FROM MEDICAL ADVICE?

Couples with an affected child
Couples who have already had a child with a birth defect or a genetic or chromosome disorder.

Couples with a family history of disease
Couples in which one or both members have a genetic disorder or a birth defect or have close relatives with one.

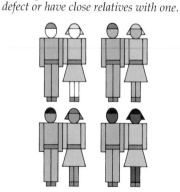

Couples from some ethnic groups
Couples from specific ethnic groups who are at increased risk for certain genetic disorders.

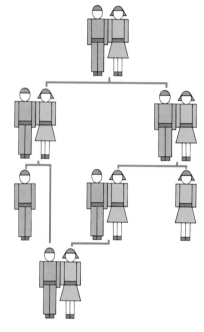

Related couples
Couples related by blood, such as a man married to his cousin's daughter.

At-risk women
Women who want to become pregnant and are 35 or older or who have had three or more miscarriages.

Doctors and genetic counselors can help individuals, couples, and families who are at risk of being affected by a genetic disorder by advising them, arranging for genetic testing, interpreting the test results, and providing treatment.

BENEFITS OF GENETIC COUNSELING

Couples who have had a child with a genetic disorder or a birth defect and are worried that a second child will also have one frequently seek genetic counseling. Such couples often believe that they are incapable of having any normal children or that they are much more likely than other couples to have another child with a birth defect or genetic disorder. These concerns are often unfounded. Not all genetic disorders are inherited – some result from new gene mutations that occur when a parent's egg or sperm is formed – and most birth defects are caused only partly by genes. In many cases, a genetic counselor can reassure concerned parents that the risk to their next child is significantly lower than they imagined.

Studying the family's health history and, in some cases, the results of blood tests of prospective parents and other family members can help genetic counselors evaluate the risk of a particular disorder. If the couple's risk is found to be increased, the counselor can explain to them the range of family planning options available (see opposite).

FAMILY PLANNING OPTIONS

All couples who are considering starting a family want to have a healthy child. Couples who are at increased risk of having a child with an abnormality may want to see a genetic counselor to discuss the various family planning options available to them.

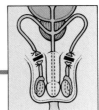
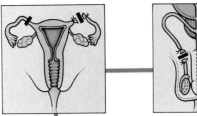

Sterilization and adoption
If a couple considers their risk of having a child with a genetic disorder to be unacceptable, one of them may choose to be sterilized. Couples who do this but who still want to have children may decide to adopt.

Determining the risk
The genetic counselor discusses with the couple their risk of having a child with a genetic disorder.

Contraception
Some couples postpone starting a family in the hope that advances in DNA testing, prenatal diagnosis, and treatment for a particular genetic disorder will increase and improve their options.

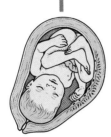

Going ahead with a pregnancy
Every couple has the right to give birth to a child, even if that child is at risk. Some couples do not wish to have prenatal tests while others choose to have some type of prenatal testing.

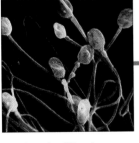

In vitro fertilization
If a woman carries a genetic abnormality, and she and her partner are determined to have a child, sperm provided by the man can be used to fertilize a donated egg in a test tube in the laboratory. A doctor then inserts the fertilized egg into the woman's uterus. In this way, a woman can give birth to a healthy baby whose genes come from her partner and a donor.

Prenatal testing
Some genetic disorders and birth defects can be detected by prenatal tests. For some abnormalities, arrangements can be made for the baby to be delivered in a hospital that specializes in treating birth defects. In other cases, a couple may choose to terminate the pregnancy.

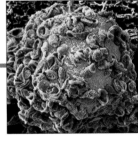

Artificial insemination
If a man has the dominant gene for a serious genetic disorder or if both members of a couple carry the recessive gene for a disorder, artificial insemination with donated sperm can be used to fertilize the woman's egg. In this technique, a doctor introduces sperm donated by a healthy man into the woman's uterus close to her time of ovulation.

BIRTH

Prenatal diagnosis

Prenatal testing (see PRENATAL TESTING AND DIAGNOSIS on page 122) often provides a couple with the reassurance that their baby will not be affected by a particular genetic disorder. However, such tests may also indicate an increased risk of an abnormality. When a serious abnormality is detected, the genetic counselor discusses with the couple the degree of mental and physical impairment that the child is likely to have; available treatment for the condition, including prenatal treatment such as surgery performed on the fetus inside the uterus; and any promising advances in treatment that may soon be available. Based on the advice they receive and their personal values, expectations, and circumstances, the couple decides on the course of action they will take. One option is abortion. Some couples would never consider abortion. However, they still may want to find out during pregnancy if there is an abnormality so that they have time to prepare themselves mentally and emotionally for their child's birth.

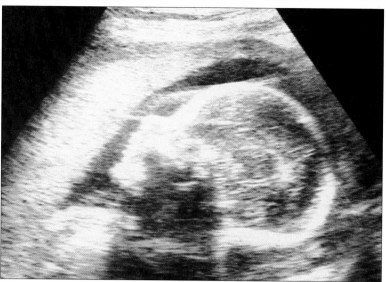

Ultrasound
The ultrasound scan above shows the head of a normal fetus. Doctors use ultrasound to monitor the development of a fetus and to diagnose skeletal abnormalities.

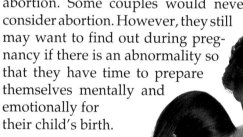

Personal choice
Some couples choose to decline any tests offered during pregnancy. Other couples want to know of an abnormality and have prenatal testing as early as possible to allow them to consider their options.

PRENATAL SCREENING

All pregnant women are offered a number of tests during pregnancy to detect abnormalities in a fetus. In most cases these tests reassure a couple that the pregnancy is progressing normally. A blood sample taken at about the 16th week of pregnancy can identify fetuses at increased risk of spine defects such as spina bifida (see page 95). Doctors use ultrasound scans to detect major physical abnormalities in a fetus.

Because of the increased risk of having a baby with Down's syndrome, pregnant women 35 and older are offered a test that analyzes a fetus's chromosomes.

SCREENING OF NEWBORNS

All newborns are screened for some inherited disorders for which early detection and treatment can improve the outlook. For example, inherited metabolic disorders such as phenylketonuria (PKU) can be identified with a simple blood test. Phenylketonuria causes mental retardation unless the affected child follows a restricted diet that is maintained throughout his or her life. All newborns in the US are tested for PKU and hypothyroidism (a disorder that can prevent normal growth and development). Some states also provide screening for metabolic disorders such as galactosemia (a disorder in which the body cannot digest a type of milk sugar).

Some genetic disorders, such as Duchenne type muscular dystrophy, can also be detected in newborns, but newborn screening for these disorders is not done because there are no available treatments for them.

DETECTING CARRIER STATUS

For a number of recessive disorders, it is possible to identify a person who carries the disease-causing gene and may be at risk of transmitting it to a child.

Ethnic groups

Members of some ethnic and racial groups are susceptible to certain recessive disorders. Tay-Sachs disease is found most frequently in Ashkenazi Jews, sickle cell anemia in blacks, thalassemia in people of Mediterranean origin, and cystic fibrosis in whites. If tests show that both members of a couple are carriers of the same abnormal recessive gene, they may want to seek genetic counseling.

Problems of screening

Although screening for genetic disorders is currently done only in specific at-risk groups (blacks for sickle cell anemia and Ashkenazi Jews for Tay-Sachs dis-

A common problem
Geneticists believe that everyone carries several defective recessive genes. Unless you have children with a person who has the same abnormal recessive gene, you are unlikely to transmit a disorder to your children.

ease), it may some day be possible to genetically screen the whole population. However, considerable controversy surrounds the idea of establishing widespread screening programs for genetic disorders. Among the major concerns are the age at which a person should be offered testing, and the impact that knowing his or her genetic susceptibilities could have on a person's employment opportunities, health insurance coverage, and choice of a partner. To be effective, screening programs must include both education and counseling.

GENE THERAPY

Medical science may soon make it possible to treat many genetic disorders that are currently untreatable. Cystic fibrosis, for example, may be controlled with regular infusions of healthy genes into an affected person's lungs. Other genetic disorders, including sickle cell anemia, may be cured by replacing defective genes with normal ones.

Diet and genetic disease
When a genetic defect makes the body unable to tolerate certain foods, a restricted diet may relieve many of the symptoms that can develop as a result. For example, people with galactosemia must exclude the milk sugar galactose from their diet.

GENETIC ANALYSIS

DOCTORS AND GENETICISTS can use a number of sophisticated laboratory techniques to study genetic diseases and help the many people who have, or are at increased risk of having, an inherited illness. Some tests can diagnose genetic diseases before birth, during childhood, or in adulthood. Other tests can determine if a person is an unaffected carrier of a specific disease-causing gene and at risk of transmitting it to a child.

Analyzing DNA
New methods of analyzing DNA have revolutionized the study and diagnosis of genetic disorders. These methods have allowed scientists to first determine the location of disease-causing genes on human chromosomes and then look for those genes in particular people.

Laboratory study of genetic disorders involves either looking for abnormalities in the body's chemistry caused by faulty genes (biochemical analysis) or looking for defects in the genetic material itself (chromosome or DNA analysis). The biggest advance in the study of genetic diseases in the last 5 years has been in DNA analysis. One of the most powerful of the new diagnostic techniques enables scientists to quickly make millions of copies of specific DNA segments. Such a huge supply of the same piece of DNA makes it possible for them to detect mistakes in individual genes more easily.

BIOCHEMICAL ANALYSIS

Biochemical testing techniques are often used to diagnose genetic disorders that are caused by deficiencies of various enzymes. These disorders are called inborn errors of metabolism (see page 71).

Using biochemical analysis, scientists can measure the levels of different substances in cells as well as the activity of specific enzymes inside cells. Tay-Sachs disease, which is caused by a deficiency or total lack of the enzyme hexosaminidase A, results in a fatal buildup of a poisonous substance in the brain. By measuring the level of activity of the enzyme in a sample of a person's blood, doctors can diagnose Tay-Sachs disease.

A problem inherent in biochemical analysis is the difficulty of obtaining body samples that contain the enzyme or other substance under study. Some proteins produced by defective genes accumulate in the blood or urine. However, many genetic defects show up only in specific cells or tissues. In these cases, a biopsy is performed to remove a small sample of tissue to make a diagnosis possible.

CHROMOSOME ANALYSIS

Chromosome analysis is used primarily to study and detect chromosome abnormalities (see page 60).

Living cells from a person, including those from a fetus, can be examined under a microscope to see if they contain

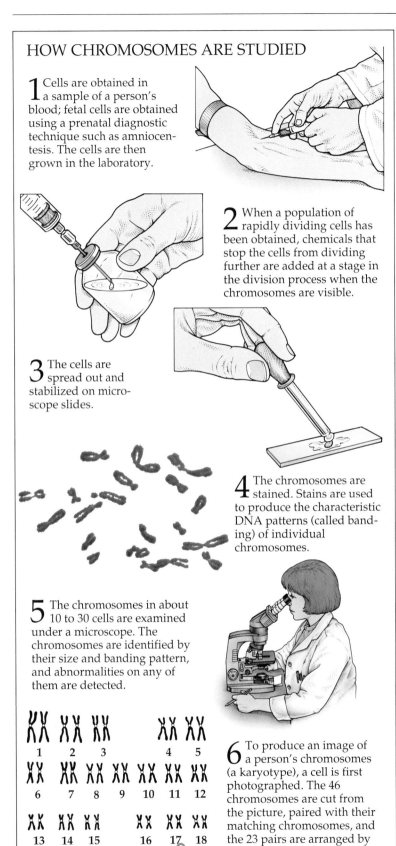

HOW CHROMOSOMES ARE STUDIED

1 Cells are obtained in a sample of a person's blood; fetal cells are obtained using a prenatal diagnostic technique such as amniocentesis. The cells are then grown in the laboratory.

2 When a population of rapidly dividing cells has been obtained, chemicals that stop the cells from dividing further are added at a stage in the division process when the chromosomes are visible.

3 The cells are spread out and stabilized on microscope slides.

4 The chromosomes are stained. Stains are used to produce the characteristic DNA patterns (called banding) of individual chromosomes.

5 The chromosomes in about 10 to 30 cells are examined under a microscope. The chromosomes are identified by their size and banding pattern, and abnormalities on any of them are detected.

1 2 3 4 5
6 7 8 9 10 11 12
13 14 15 16 17 18
19 20 21 22 X X Y

6 To produce an image of a person's chromosomes (a karyotype), a cell is first photographed. The 46 chromosomes are cut from the picture, paired with their matching chromosomes, and the 23 pairs are arranged by number. The karyotype at left is of a person with an extra X sex chromosome.

any chromosome abnormalities. For example, tests can detect the presence of an extra chromosome 21 in a fetus's cells, which is the most common cause of Down's syndrome. In a couple with a history of miscarriages, tests can determine whether either partner has a rearrangement of chromosomes such as a balanced translocation (see page 64). Such a genetic rearrangement can increase the couple's risk of having a child with a chromosome disorder.

Blood is the most common source of cells for study. White blood cells are easy to obtain and they divide quickly (within 48 hours) in the laboratory. For prenatal diagnosis, the most common source of fetal cells is amniotic fluid obtained by amniocentesis (see page 126). It takes an average of 2 to 3 weeks to grow enough fetal cells in the laboratory to study.

DNA ANALYSIS

DNA analysis has been used primarily to identify and study single-gene disorders (see page 70), such as hemophilia and sickle cell anemia. Studying DNA enables doctors to determine if a person is a carrier of a specific X-linked recessive or autosomal (non-sex-linked) recessive disorder. DNA analysis is also used to diagnose many common single-gene disorders in a fetus.

Using genetic probes

Most DNA analysis involves cutting a sample of a person's DNA into millions of pieces, using biological "scissors" called restriction enzymes. The pieces are then examined using other pieces of DNA called genetic probes. Genetic probes can directly detect, for example, a type of gene mutation called a deletion that involves the loss of part of a gene. Approximately 65 percent of boys with Duchenne type muscular dystrophy are missing parts of a particular gene on the X chromosome. Genetic probes can detect this gene defect in a fetus before birth.

WHAT IS A GENETIC PROBE?

A genetic probe is a piece of single-stranded human DNA that has been radioactively labeled. A probe contains a DNA pattern that matches the pattern in the DNA of the person being studied. Some probes can directly detect the presence or absence of specific genes (both normal and abnormal). Other genetic probes are used to identify so-called marker DNA patterns that serve as landmarks for particular disease-causing genes.

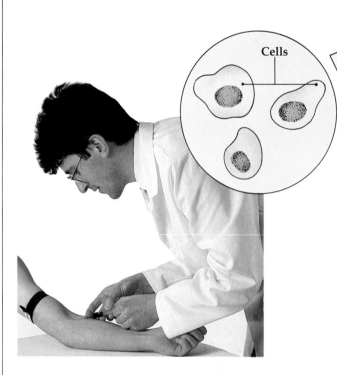

Cells

Mass of DNA in test tube

1 DNA is obtained from white blood cells in a sample of a person's blood or from fetal cells obtained by amniocentesis or chorionic villus sampling.

DNA base sequence to be studied

Film being examined

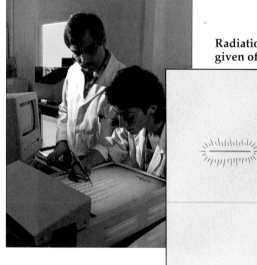

Radiation given off

Gene mutations

The illustration shows the results of DNA tests on four boys using genetic probes to search for parts of the Duchenne type muscular dystrophy gene. Each dark band represents radiation given off when a probe binds to a matching DNA fragment. Column D shows five bands in the DNA of a healthy boy. The other columns represent DNA from boys with the disorder. The absence of some bands or the appearance of bands of a different size indicate mutations in the disease-causing gene.

A   B   C   D

6 If the probe encounters the matching DNA pattern, the probe will bind to that piece of DNA. This binding appears on a film called an autoradiograph as a dark band at a particular location. If the DNA pattern being sought is not present, the probe does not bind, no radiation is given off, and no dark band appears on the film.

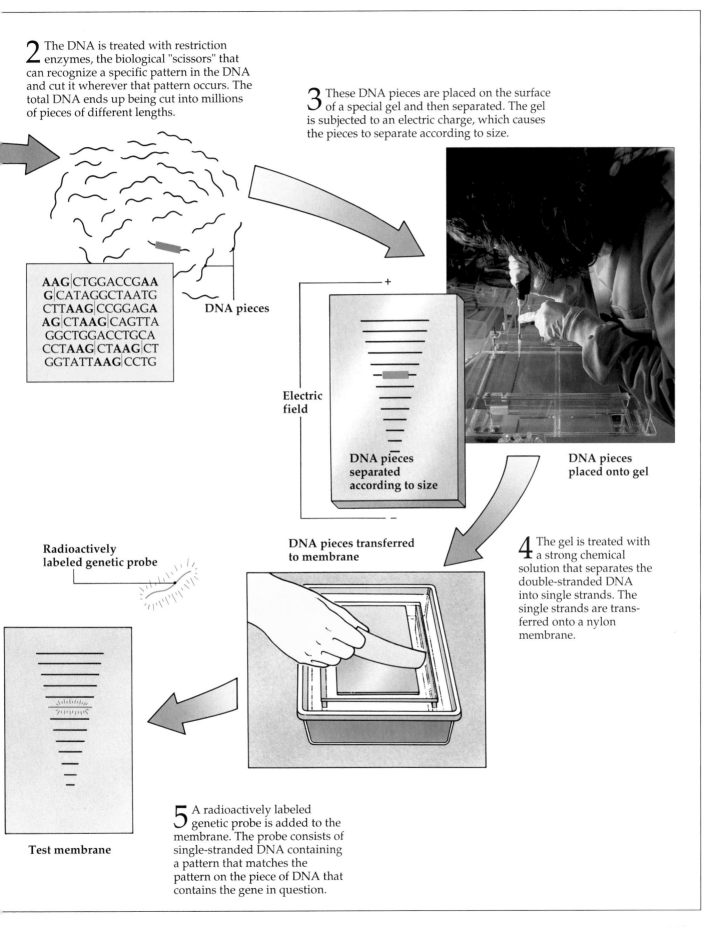

2 The DNA is treated with restriction enzymes, the biological "scissors" that can recognize a specific pattern in the DNA and cut it wherever that pattern occurs. The total DNA ends up being cut into millions of pieces of different lengths.

3 These DNA pieces are placed on the surface of a special gel and then separated. The gel is subjected to an electric charge, which causes the pieces to separate according to size.

AAG|CTGGACCGAA
G|CATAGGCTAATG
CTT**AAG**|CCGGAGA
AG|C**TAAG**|CAGTTA
GGCTGGACCTGCA
CC**TAAG**|C**TAAG**|CT
GGTATT**AAG**|CCTG

DNA pieces

+

Electric field

DNA pieces separated according to size

−

DNA pieces placed onto gel

DNA pieces transferred to membrane

4 The gel is treated with a strong chemical solution that separates the double-stranded DNA into single strands. The single strands are transferred onto a nylon membrane.

Radioactively labeled genetic probe

Test membrane

5 A radioactively labeled genetic probe is added to the membrane. The probe consists of single-stranded DNA containing a pattern that matches the pattern on the piece of DNA that contains the gene in question.

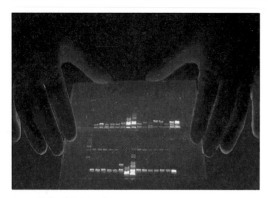

Amplified DNA fragments
The photograph above shows a scientist examining DNA pieces that have been copied many times (amplified) and separated in a gel. The pieces are visible after being stained under ultraviolet light.

Amplifying DNA

A major breakthrough in DNA technology – called the polymerase chain reaction (PCR) – significantly improves the way in which genetic disorders are diagnosed. In PCR, millions of copies of a piece of a person's DNA are produced quickly (in a process called amplification), enabling scientists to home in on a disease-causing gene. In other words, instead of having to look for a gene in a "haystack," PCR provides geneticists with a whole haystack full of the same genes. This enables them to locate and isolate specific DNA pieces, even if they are buried in large amounts of other DNA. Individual pieces can be identified by cutting, separating, and staining. Staining helps to detect the presence or absence of specific pieces of DNA and provides geneticists with information about specific genes. PCR allows a sample of DNA to be tested in a few hours, compared with the several days required using genetic probes.

DNA markers

Scientists have discovered that, just as blood type and hair color differ, human DNA comes in a wide variety of detectable patterns. This diversity in our DNA provides doctors with useful reference points, or markers, to help them diagnose genetic diseases in families. When geneticists study the DNA of some families known to be affected by a specific disease, they find that a particular DNA marker and the disease gene are almost always transmitted together from parent to child. In such cases, the marker and the disease gene must be very close to each other on the same chromosome. When scientists cannot determine the exact location of an individual gene on a chromosome, they use DNA markers to point the way. The genes for many of the major single-gene disorders, including cystic fibrosis and sickle cell anemia, have been identified using DNA markers.

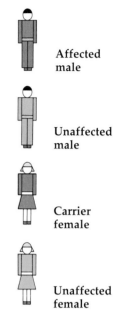

Tracking DNA markers
The illustration below shows the results from testing eight members of a family for two forms of a DNA marker linked to a specific X-linked recessive disorder. Form 1 is the marker for the normal gene; form 2 is the marker for the disease gene. Affected members (males) have form 2 of the marker on their only X chromosome. Women B and D have both forms 1 and 2, making them unaffected carriers. Because woman H has only form 1 (presumably on both of her X chromosomes), she is unlikely to be a carrier.

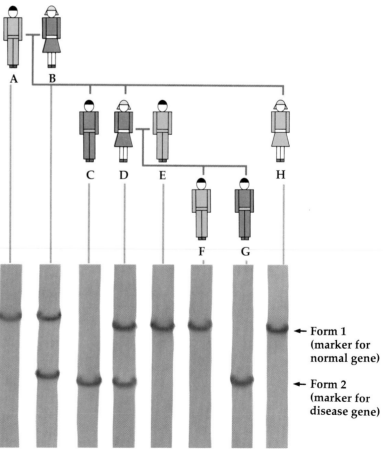

DNA FINGERPRINTING

Almost all DNA markers exist in more than one form. Like the bar codes used in supermarkets to identify grocery items, DNA markers come in many different combinations. These combinations form patterns that are unique for every person except an identical twin (whose twin has identical DNA). This pattern is referred to as a person's DNA fingerprint. Half of the bands in your DNA fingerprint come from each of your parents. For this reason, DNA fingerprints are used in paternity cases to determine, with high precision, whether a man is a child's biological father, as shown below. DNA fingerprints are also being used increasingly to help solve crimes.

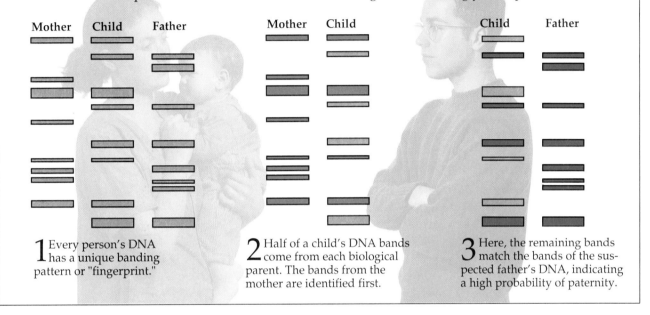

1 Every person's DNA has a unique banding pattern or "fingerprint."

2 Half of a child's DNA bands come from each biological parent. The bands from the mother are identified first.

3 Here, the remaining bands match the bands of the suspected father's DNA, indicating a high probability of paternity.

Once it has been established that a specific DNA marker is located very close to a particular abnormal gene, that marker (and others similar to it) can be used to determine whether or not a person carries the gene. This technique is extremely useful in identifying carriers of many genetic disorders and in diagnosing genetic diseases during pregnancy.

Markers are often used to establish the transmission pattern of a specific disease gene in a family. Analysis of the transmission pattern is called gene tracking. The accuracy of these tests depends on how close to each other the DNA marker and the gene are on a chromosome.

Detecting gene mutations

Once the location of a defective gene on a chromosome is known, a technique called DNA sequencing (see right) helps scientists understand the nature of the gene's mutation. For some diseases, such as cystic fibrosis, the gene mutations have been identified. Once a mutation has been recognized, analysis of the DNA of people at risk can determine whether or not they have it. However, if the number of different gene mutations that can cause a disease is too large, direct DNA analysis of this type is not of practical use.

DNA sequencing
The scientist at left is reading the chemical code of a defective gene and comparing that code to the code of a normal gene. Once specific gene mutations have been identified, they can be used to help geneticists diagnose genetic disorders and identify carriers of those disorders.

GENETIC COUNSELING

GENETIC COUNSELING can benefit people who want to have children but who think they may be at increased risk of having a child with a genetic disorder or a birth defect. Genetic counseling can help a person who is affected, or whose family is affected, by a genetic disorder and who wants to learn more about it. Genetic counseling is also valuable in identifying potentially affected members of a family and encouraging them to be examined.

Getting help
Genetic counseling can help people who may be at risk of transmitting a genetic disorder to their children. Genetic counseling can also help couples who simply want to know more about a disorder that runs in one of their families or that has appeared unexpectedly in one of their children.

If a couple already have a child with a genetic disorder or a birth defect, a genetic counselor can explain to them the known causes of the particular abnormality and how it is likely to affect their child. The genetic counselor evaluates the couple's risk of having another child with the same disorder or defect and discusses the various family planning options that are available to them (see page 105).

WHO SEEKS COUNSELING?

People are referred for genetic counseling for a variety of reasons. In addition to couples who already have an affected child, many prospective parents have experienced repeated miscarriages or are concerned about a genetic disorder or other risk factor in their families.

An affected child
Counseling is essential for parents who have a child with a genetic disorder or a birth defect. If the child's disorder has been identified, the parents may want to know the risk of the disorder or defect occurring in other children they may have or in the children of other family members. In cases in which a diagnosis has not been made, the parents may choose to seek further advice from a doctor who specializes in genetics.

Repeated miscarriages
Miscarriage is common. One in eight pregnancies ends in miscarriage at an early stage; about half of early miscarriages are caused by chromosome abnormalities in the fetus. Most of these genetic errors are not inherited – they occur randomly during the formation of a parent's egg or sperm and usually only happen to a couple once. However, if a woman has three or more miscarriages, her doctor might recommend that she and her partner have chromosome analysis and genetic counseling. In about

5 percent of these couples, tests reveal an abnormal chromosome rearrangement (see page 64) that can increase their risk of having another miscarriage or of having a child with a birth defect. But some couples who have normal chromosomes themselves have repeated miscarriages caused by chromosome abnormalities in the fetus. Prenatal diagnosis is recommended for such couples.

An affected parent

A person who was born with a genetic disorder, or who developed one later in life, may want to know the risk of transmitting the disorder to his or her children. If the disorder is serious, a couple may choose to consider various family planning options (see page 105), such as adoption or artificial insemination with a donor's sperm (if the man is af-

fected). Some couples decide not to have children, while others decide, after experiencing a particular disorder and its effects themselves, that it is not sufficiently serious to prevent them from having children. For example, deaf people are often willing to take the chance that their children might also be deaf. However, before making a final decision, they may want to know if other abnormalities that they consider more serious can also be passed on to offspring.

Family history

Some couples are reluctant to have children because they have a family history of a genetic disorder, or because they

Looking at a family history
Mike and Susan want counseling because Mike's brother John has Friedreich's ataxia, an incurable disorder that occurs when a person inherits two copies of a recessive disease gene. They are worried because they have a common great-grandparent. Mike has a 2 in 3 risk of being a carrier and Susan is at increased risk of also being a carrier. DNA testing can help them get a clearer picture of their risks.

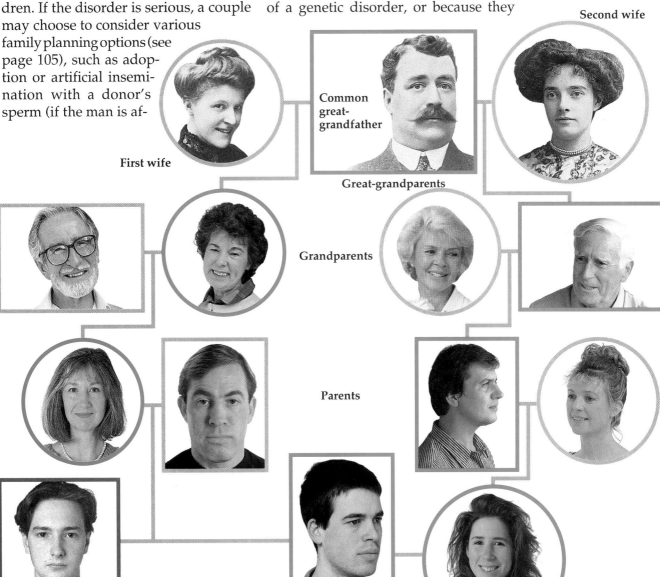

First wife

Common great-grandfather

Second wife

Great-grandparents

Grandparents

Parents

John

Mike

Susan

have a relative with a disorder that they think is hereditary. Genetic counseling is especially valuable for such couples because the counselor can provide them with all the known facts about the disorder in question. Couples often find that the risk is much lower than they had imagined.

Risk factors

Some couples seek counseling even though they have no family history of a genetic disorder and are not affected themselves. Close relatives, such as first cousins, who plan to marry may be concerned about having a child with an inherited disorder. Women 35 and older may be worried about disorders such as Down's syndrome that are known to be more common in children of older women. Because the genes for some recessive disorders are more common in certain ethnic groups, people belonging to those groups are at increased risk of being

carriers. For example, the gene that causes the metabolic disorder Tay-Sachs disease is most common in Ashkenazi Jews. The incidence of the disease in this group used to be about one in 2,500, which was 100 times higher than in any other group. However, since the introduction of carrier testing, genetic counseling, and prenatal testing in the US, the incidence has declined significantly.

WHO PROVIDES COUNSELING?

Genetic counseling is usually offered by specialized clinics that are associated with hospitals. The staff at a genetics clinic usually includes a medical geneticist (a doctor who has expertise in diagnosing genetic disorders), a genetic counselor (trained in the scientific and practical aspects of genetics as well as counseling techniques), and scientists who perform laboratory tests to confirm diagnoses and detect carrier status. A doctor from a clinic that specializes in treating a specific disease, such as cystic fibrosis, may also provide counseling.

Pregnancy after age 35
The incidence of some chromosome disorders, particularly Down's syndrome, increases significantly in children of women over the age of 35. A woman 35 or older who becomes pregnant, or plans to have a child, may want to seek genetic counseling to discuss her risks and the options available to her.

Groups at risk
Couples in some ethnic or racial groups who are at increased risk of being carriers of specific genetic disorders may want to seek genetic counseling to discuss their childbearing options. For example, one in 10 black Americans carries the gene for sickle cell anemia but is personally unaffected. If two carriers have children, each child has a 1 in 4 chance of inheriting two copies of the gene and, therefore, of having the disorder.

CASE HISTORY
A MISSING ENZYME

RACHEL AND MICHAEL **were deeply distressed when their second child David died when he was 3. The family doctor told the couple that their son's death was caused by a hereditary disease. Although their daughter Deborah seemed to be perfectly healthy, the doctor recommended that they seek genetic counseling before deciding to have more children.**

PERSONAL DETAILS
Names Rachel and Michael Cohen
Ages Both 26
Occupations Rachel is a lawyer; Michael is a dentist.
Family Both Rachel and Michael are Ashkenazi Jews. Their daughter Deborah is 5 years old.

MEDICAL BACKGROUND
Until the age of 6 months, the couple's son David had been a very healthy baby. Then he became blind, deaf, and paralyzed and began to deteriorate mentally and physically. A series of tests showed that David had the fatal inherited metabolic disorder Tay-Sachs disease, which affects the nervous system. David died when he was 3. Rachel and Michael and their daughter Deborah are all healthy and no other relatives have had the disorder.

GENETIC COUNSELING
The genetic counselor explains that Tay-Sachs disease is caused by the absence of an enzyme called hexosaminidase A. She explains that, among Ashkenazi Jews, one in 30 (compared with one in 300 in the general population) carries the abnormal gene that blocks production of the vital enzyme. Tay-Sachs disease follows an autosomal recessive pattern of inheritance, which means that David must have inherited a copy of the recessive disease-causing gene from both of his parents. Because Rachel and Michael are carriers, each of their children has a 25 percent chance of inheriting two copies of the defective gene and, therefore, of having Tay-Sachs disease. The counselor explains that, even though the couple already has had one affected child, it does not mean that the risk to their future children is any less. She tells them that it is possible to diagnose the disease during pregnancy from tests on fetal cells. She reminds Rachel and Michael that there is presently no treatment for Tay-Sachs disease and that it is always fatal. The counselor explains that each of their unaffected children has a 2 in 3 chance of being a healthy carrier of the disease gene and so is at risk of transmitting it to his or her own offspring. Because their daughter Deborah may be a carrier of the abnormal gene, she may decide to be tested when she is older and considering having children. The counselor recommends that the couple's brothers and sisters be offered testing to determine if they are also carriers. If they are found to be carriers, they may be interested in seeking genetic counseling as well.

THE OUTCOME
During the next year, Rachel and Michael see the genetic counselor several times before making a decision about having another child. When Rachel becomes pregnant 18 months after the initial consultation, prenatal diagnosis shows that the fetus does not have Tay-Sachs disease. Rachel and Michael now have two healthy daughters.

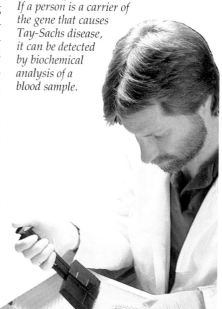

Detecting carrier status
If a person is a carrier of the gene that causes Tay-Sachs disease, it can be detected by biochemical analysis of a blood sample.

PRELIMINARY INVESTIGATIONS

A genetic counselor does extensive research before providing advice. First, he or she usually draws up a detailed family health tree that highlights all members of the family who have had a certain disorder, and those who are either obvious or possible carriers (see page 82). Medical records of affected relatives may provide information about noticeable signs and symptoms of the disorder and help to identify the effects of the disorder on different family members. Affected relatives and, in some cases, relatives who are at risk may be asked to come to the clinic for a physical examination. The examination focuses on specific body systems, depending on the disorder. The examination often includes a careful search for signs, such as unusual markings on the skin and characteristic facial features, that are not in themselves abnormal but that might help identify people who have the disorder or who are carriers. After the physical examination, the medical geneticist may recommend more testing, such as DNA analysis or biochemical analysis (see GENETIC ANALYSIS on page 108), to confirm a diagnosis.

Evaluating the risks

After gathering and analyzing the information, a genetic counselor can usually evaluate the risks to the person being counseled. Those risks may include the person's risk of developing a particular condition and of transmitting it to children. For some disorders, the risks can be accurately determined by laboratory tests. For example, DNA testing can confirm whether a person is a carrier of the gene for cystic

Reviewing medical records
If a person seeks counseling because of a family history of a particular genetic disorder, the medical records of affected relatives may be used to chart the course of the disease in the family and study its pattern of inheritance.

Physical examination
Examination of the person at right has revealed many coffee-colored skin patches – a characteristic of the inherited disorder neurofibromatosis. This person would be advised that he or she – and any child who inherits the abnormal gene – may be only mildly affected or may develop more serious symptoms, such as skin tumors or nervous system abnormalities.

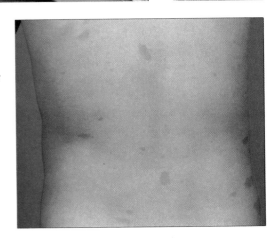

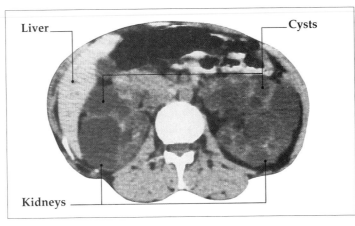

Liver — Cysts

Kidneys —

Adult polycystic kidney disease
The computed tomography (CT) scan at left shows multiple cysts in the kidneys of a person with adult polycystic kidney disease. The progressive disorder usually causes no symptoms until the person has kidney failure later in life. In children of affected people, a DNA linkage test can help to predict if they have the abnormal gene.

fibrosis. For many other disorders, DNA testing is not yet available and a person's risk must be estimated on the basis of his or her family history and statistics compiled from studies of other families affected by the disorder.

WHAT WILL BE DISCUSSED?

A genetic counselor can provide the known facts about a particular disorder, including its cause, the hereditary and environmental influences that may play a role, its pattern of inheritance, its probable course, and the availability of treatment. This information-gathering part of genetic counseling is crucial – by helping couples to understand how a disorder might affect them and their children, it enables them to make informed decisions about having a family.

Risks

The counselor explains to a couple the estimated risks to their future children and to any children that those children may have. Many people find the concept of risk a difficult one to deal with. A risk that one couple finds prohibitively high may be acceptable to another couple. Genetic counselors are trained to help families understand the implications of those risks. For example, a misconception that many people have is that once a couple has a child with a genetic disorder, their risk of having another child with the same disorder is reduced. This is not the case – the risks calculated for a person or couple are the same for each pregnancy, regardless of the health of any previous children. A 1 in 4 risk for a particular disorder does not mean that if you have four children one will be affected and the other three will be unaffected; it means that each child that you have has the same 25 percent risk of inheriting the disorder.

Available options

The next stage in counseling is a discussion of the options available to a couple to help them deal with the possible risks. Options may include prenatal testing to determine if the fetus has the disorder and alternative methods of having children (see FAMILY PLANNING OPTIONS on page 105). Some people are concerned that their genetic counselor will try to persuade them to follow a particular course of action. They need not worry – counselors specifically avoid expressing their own opinion. People are always reminded that the final decision is theirs.

In addition to providing an informative explanation of the choices a couple can make, the counselor supports them in their decision-making. Each couple must make the decision that

Explaining the risks
Counselors explain to a couple that the evaluation of risks for a particular disorder is the same for every pregnancy, regardless of the health of any children the couple already has.

ASK YOUR DOCTOR
GENETIC COUNSELING

Q My sister had cystic fibrosis and died when I was 7. If I am a carrier, can I pass the disorder on to my children?

A A child with cystic fibrosis has inherited the gene from both parents. DNA analysis can help determine whether or not you are a carrier; your risk is about 2 in 3. However, your children are only at risk if your partner is also a carrier (a 1 in 20 chance among whites) and if they inherit the gene from both of you.

Q I am 20 years old and my father has Huntington's chorea. I am especially worried because I look like my father. Does that mean I am more likely to have inherited the disease from him?

A No. Because the gene that causes Huntington's chorea is dominant, each of your father's children has the same 50 percent chance of inheriting it. The genes that determine physical characteristics are separate from the gene that causes Huntington's chorea. Therefore, your resemblance to your father has no influence on whether or not you inherited the disease gene.

Q We have a mildly retarded son and two healthy daughters. Our doctor thinks that our son's disorder might be genetic and recommends counseling. But, since my husband has had a vasectomy, is there any reason to do so?

A Yes. A genetic counselor may be able to diagnose your son's disorder and explain its effects and the ways in which it can be treated. The counselor can also discuss the risk of any of your children transmitting the disorder to their offspring.

they feel is best for them and their family. Usually the final decision is based not only on the medical facts about a disorder and the degree of risk, but also on the couple's own experience with the disorder and their feelings and expectations at the time of counseling. Decision-making can be particularly difficult when the disorder has varying degrees of severity or when family members disagree about a course of action. Genetic counselors know that there are no right or wrong decisions – their job is to help families evaluate the facts.

Psychological effects

Many people find it painfully difficult to cope with the news that a child or another family member has a serious hereditary disorder. Counselors emphasize that there is no reason for a person to feel guilty or at fault for carrying a specific genetic trait. Genetic counselors can help people to better understand and deal with the impact of a diagnosis and the risk of a disorder occurring again in their family. In some cases, genetic counselors may refer a person to a psychotherapist or to a support group.

Candid discussion
Many couples find that a complete and open discussion about a genetic disorder or birth defect helps alleviate concerns that they may have had for a long time. This thorough discussion is one of the most important goals of genetic counseling.

CASE HISTORY
WEAK MUSCLES

THREE-YEAR-OLD RICHARD has developed much more slowly than other children his age. His parents are also concerned about his difficulty walking and climbing stairs, and his constant complaints of being tired. After his third birthday, Richard showed no signs of improvement and his parents decided to seek advice from their doctor.

PERSONAL DETAILS
Name Richard Adams
Age 3
Family Richard's parents are both healthy. He has no brothers or sisters.

MEDICAL BACKGROUND
Richard was born after a normal pregnancy. He was slow in learning to sit up and walk and did not play actively with children his age.

THE CONSULTATION
While examining Richard, the doctor finds that the boy's muscles are weak. After Richard's parents describe the boy's symptoms, the doctor orders several tests, including a blood test and a muscle biopsy.

THE DIAGNOSIS
The results of the tests confirm that Richard has DUCHENNE TYPE MUSCULAR DYSTROPHY, an inherited muscle disorder caused by the lack of a protein called dystrophin. The doctor explains that Richard's muscles will continue to deteriorate and that eventually he will need to use a wheelchair. He also tells the couple that, because there is no treatment or cure, children with the disorder rarely survive their teens – they usu-

ally die of respiratory infections or respiratory failure when muscles in their diaphragm become affected. However, the doctor also informs them that researchers are currently looking for a way to produce the missing protein, dystrophin, in people who lack it. He recommends that the couple see a genetic counselor to learn more about the disorder.

GENETIC COUNSELING
The genetic counselor explains to Richard's parents that Duchenne type muscular dystrophy is caused by an abnormal recessive gene on the X sex chromosome. The mothers of affected males are usually unaffected carriers of the gene. Because Richard has the disorder, his mother Sarah is probably a carrier of the defective gene and she in turn may have inherited it from her carrier mother. The counselor arranges for DNA tests to be performed on Sarah, her mother, and her two sisters, to

determine if they are carriers. She explains further that, if a female carrier has children, each of her sons has a 50 percent chance of having the disorder; each daughter has a 50 percent chance of being a carrier.

THE OUTCOME
The results of the blood tests confirm that Sarah and her mother are carriers. Sarah's two sisters are not carriers, which means that they inherited the normal copy of the gene from their mother. The counselor explains to Sarah that the disorder can be detected by prenatal testing if she becomes pregnant again. Sarah and her husband discuss the risks and decide to spend their time providing Richard with the special care that he needs before they make a decision about having more children.

Locating the disease gene
The counselor explains to the couple the results of Richard's DNA analysis, which confirms that he has the gene for Duchenne type muscular dystrophy. She tells them that the same test can be performed early in pregnancy if Sarah should become pregnant again.

PRENATAL TESTING AND DIAGNOSIS

U NTIL RECENTLY, couples had no way of knowing if their baby would be born healthy or with a serious genetic disorder or birth defect. But now, for many abnormalities, prospective parents can choose from a number of new diagnostic tests to learn long before birth whether or not a fetus is affected. Such information can help them make important decisions about the pregnancy.

Prenatal testing
Many couples decide to have prenatal testing to find out early in a pregnancy if the fetus has an abnormality. Although some birth defects can be treated or corrected by surgery, many cannot.

A variety of tests can be performed during pregnancy to detect abnormalities in a fetus. Although many couples decide to have prenatal tests, some at-risk couples choose to go through a pregnancy without any testing at all. A couple is informed of the prenatal tests that may be helpful in their case. If the fetus has been diagnosed with a genetic disorder or birth defect, couples are often referred for genetic counseling (see GENETIC COUNSELING on page 114).

RISK OF MISCARRIAGE ASSOCIATED WITH PRENATAL TESTING

Miscarriage is the major risk associated with prenatal testing techniques. Couples considering having prenatal tests should discuss the risks of the procedure and the importance to them of knowing the results.

TECHNIQUE	RISK OF MISCARRIAGE
Amniocentesis	Increased risk of 0.5 in 100
Chorionic villus sampling	Increased risk of 2 in 100
Fetal blood sampling	Increased risk of 2 in 100
Fetoscopy	Increased risk of 3 to 6 in 100
Maternal blood sampling	No increased risk
No test	It is estimated that as many as 20 percent of all pregnancies end in miscarriage. Most miscarriages occur in the first 10 weeks of pregnancy.
Ultrasound scanning	No increased risk

PRENATAL TESTING

Couples who are at increased risk of having a child with a genetic disorder or birth defect include those with a family history of a disorder, couples who already have one affected child, and pregnant women who will be 35 or older when they deliver. In addition, pregnant women are regularly offered tests to detect some disorders such as spina bifida (in which the fetus's spine fails to develop normally). The procedures used in prenatal diagnosis include ultrasound scans to monitor fetal development, blood tests for both the pregnant woman and the fetus, and amniocentesis and chorionic villus sampling, which are used to obtain fetal cells. Fetal cells are studied using specialized laboratory tests such as biochemical analysis (which measures the level of activity of specific proteins), chromosome analysis, and DNA analysis (see GENETIC ANALYSIS on page 108). Most couples who decide to have prenatal testing are reassured by the test results that there are no problems.

SCREENING PROGRAMS

The only major prenatal screening program in the US today is alpha-fetoprotein screening, which is used to detect neural tube defects such as spina bifida (see page 95). A new screening procedure now under development for Down's syndrome may eventually be made available to all women regardless of their age. Currently, only women 35 and older are regularly offered prenatal testing for Down's syndrome. Doctors are increasingly using ultrasound scanning to detect structural abnormalities in fetuses.

Blood tests in pregnancy
As soon as a pregnancy is confirmed, a sample of the woman's blood is sent to a laboratory for analysis. Besides determining blood count and blood type, the blood tests can reveal signs of infection with organisms such as the hepatitis B virus, the rubella virus, and the bacterium that causes syphilis. A woman may also choose to be tested for the AIDS virus (HIV).

Human immunodeficiency virus (HIV)

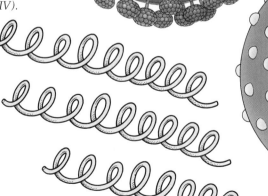

BENEFITS OF TESTING
The potential benefits of testing before and/or during pregnancy include the following:
◆ Couples whose test results show an abnormality can seek genetic counseling to discuss their family planning options.
◆ Couples with normal test results are reassured.
◆ At-risk couples can start a pregnancy knowing that, if tests show that the fetus has a serious defect, they can prepare themselves emotionally for their baby's birth or they can choose to terminate the pregnancy.
◆ Hospital preparations can be made to immediately care for a newborn with a birth defect.

Hepatitis B virus

Rubella virus
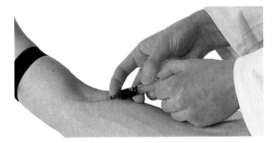

Syphilis bacteria

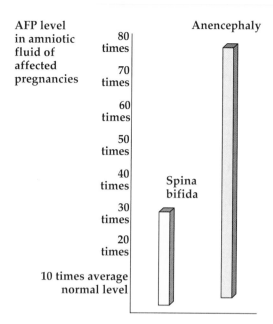

AFP level in amniotic fluid of affected pregnancies

Anencephaly

80 times
70 times
60 times
50 times
40 times
30 times
20 times

Spina bifida

10 times average normal level

Measuring alpha-fetoprotein

The graph at left compares the range of alpha-fetoprotein (AFP) levels found in pregnancies in which the fetus has the neural tube defects spina bifida (open spine) or anencephaly (in which the brain is missing) with the average normal level. When fetuses have spina bifida, AFP levels in amniotic fluid are raised between two and 30 times the average. When fetuses have anencephaly, AFP levels in amniotic fluid are raised between five and 80 times the average.

Alpha-fetoprotein screening

More than 95 percent of children with neural tube defects (in which the brain or spine fail to develop normally) are born to couples who have no family history of such abnormalities. Most affected pregnancies can be found by screening pregnant women with a simple blood test. Alpha-fetoprotein is a protein made by the liver that is present in the blood and body fluids of a developing fetus. During pregnancy, some alpha-fetoprotein passes from the fetus into the woman's bloodstream; the amount can be accurately measured at about 16 weeks of pregnancy. If the fetus has a spinal malformation, an increased amount of alpha-fetoprotein leaks into the amniotic fluid and passes into the woman's bloodstream. If the level is high, a second sample may be taken. If the level in this sample is also high, the doctor orders an ultrasound scan or amniocentesis (or sometimes both) to check for a fetal malformation. A pregnant woman's alpha-fetoprotein level can be higher than normal for reasons other than a malformation (such as a multiple pregnancy) or for no reason. A lower-than-normal level appears to indicate an increased risk of a fetus with Down's syndrome.

Chromosome abnormalities

Each year in the US, more than 20,000 infants are born with chromosome defects (see CHROMOSOME ABNORMALITIES on page 60). All known chromosome abnormalities can be diagnosed by prenatal chromosome analysis (see page 108) of fetal cells. However, the procedures used to obtain cells from the fetus – amniocentesis and chorionic villus sampling – carry an increased risk of miscarriage. For this reason, they are performed mostly on women who are at increased risk, such as those who will be 35 or older at the time of delivery or women whose level of alpha-fetoprotein is shown to be reduced. With the availability of alpha-fetoprotein screening, it is now possible to identify more women (even those under age 35) who may be at increased risk of having a baby with Down's syndrome. That risk is

Testing younger women
Women under age 35 who are found from blood tests to be at increased risk of having a baby with Down's syndrome may have amniocentesis to see if the fetus is affected. In contrast, if the blood test shows that a woman 35 or older is at low risk of having a baby with Down's syndrome, she may choose not to have amniocentesis, which carries some risk of miscarriage.

evaluated from a combination of factors, including a woman's age and the levels of alpha-fetoprotein and other biochemical indicators in her blood. In this way, women found to be at increased risk may choose to have amniocentesis, regardless of their age (see TESTING YOUNGER WOMEN on page 124).

PREGNANCIES AT RISK

A couple who already have an affected child or who have a family history of a genetic disorder or birth defect may be at increased risk of having another affected child. For some inherited disorders, family studies using DNA analysis may help doctors evaluate a couple's risks and help them decide whether the risks associated with prenatal testing procedures such as amniocentesis are acceptable to them. If a pregnancy is found to be at increased risk, a doctor

may recommend that procedures such as ultrasound scanning, amniocentesis, or fetal blood sampling be performed, depending on the suspected abnormality. In most cases, tests show no defects. If one is detected, treatment is sometimes possible, either during pregnancy or at birth. However, many birth defects are untreatable. For pregnancies in which a serious untreatable disorder has been found, couples often seek genetic counseling to learn about their options.

ULTRASOUND SCANNING

Ultrasound scanning is a prenatal diagnostic technique used to help doctors monitor the growth and development of a fetus. Like the underwater sonar used to detect submarines, ultrasound uses high-frequency sound waves (which are passed through the pregnant woman's abdomen) to see the fetus. The reflected echoes provide an image of the fetus and

DETERMINING SEX OF A FETUS

The sex of a fetus can be determined either by chromosome analysis or DNA analysis. Usually these tests are performed to determine sex only if a woman is at risk of transmitting an X-linked recessive disorder. These disorders mostly affect boys because boys have only one X chromosome, which, if defective, causes illness. If the fetus is found to be male, more tests are performed. If the fetus is female, the woman is reassured that her pregnancy is not at risk.

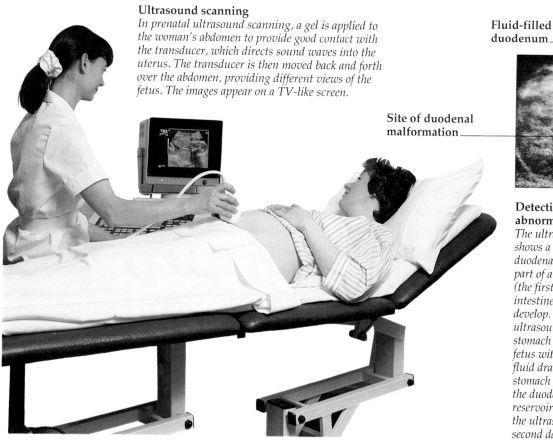

Ultrasound scanning
In prenatal ultrasound scanning, a gel is applied to the woman's abdomen to provide good contact with the transducer, which directs sound waves into the uterus. The transducer is then moved back and forth over the abdomen, providing different views of the fetus. The images appear on a TV-like screen.

Stomach
Fluid-filled duodenum
Site of duodenal malformation

Detecting structural abnormalities
The ultrasound scan above shows a disorder called duodenal atresia, in which part of a fetus's duodenum (the first part of the small intestine) has failed to develop. In a normal ultrasound scan, only the stomach is visible. But, in a fetus with duodenal atresia, fluid draining from the stomach gets trapped in the duodenum, forming a reservoir that appears in the ultrasound scan as a second dark area.

placenta, which is displayed on a TV-like monitor. In women known to be at increased risk, ultrasound scanning may be used to help identify structural abnormalities for which there are no specific laboratory methods of diagnosis. For example, ultrasound scans can detect unusually short bones that are characteristic of achondroplasia (dwarfism) and some other skeletal abnormalities. Ultrasound scanning can also be used to diagnose abnormalities such as neural tube defects and heart, kidney, bladder, and gastrointestinal disorders that are present before birth. Ultrasound, which is considered to be risk-free, is usually performed at 16 to 18 weeks of pregnancy; if abnormalities are found, repeat scans and amniocentesis are often performed.

AMNIOCENTESIS

Amniocentesis is a prenatal diagnostic procedure that involves inserting a thin needle through a woman's abdomen into the uterus to remove a sample of amniotic fluid, which surrounds the fetus. Amniotic fluid contains cells and chemicals from the fetus and can be studied in a number of ways for abnormalities. Amniocentesis is usually performed at about the 16th week of pregnancy. The increased risk of miscarriage associated with the procedure is about 0.5 in 100. Ultrasound scanning (see page 125) is

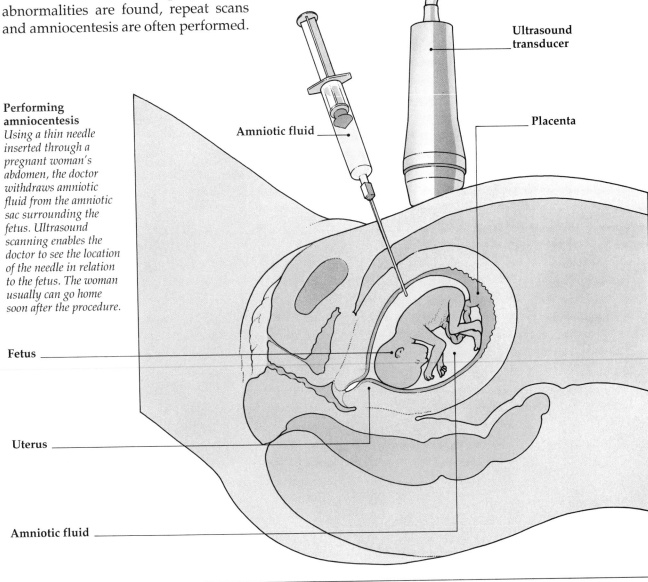

Performing amniocentesis
Using a thin needle inserted through a pregnant woman's abdomen, the doctor withdraws amniotic fluid from the amniotic sac surrounding the fetus. Ultrasound scanning enables the doctor to see the location of the needle in relation to the fetus. The woman usually can go home soon after the procedure.

Ultrasound transducer

Amniotic fluid

Placenta

Fetus

Uterus

Amniotic fluid

used simultaneously to monitor the position of the fetus and to guide the insertion of the needle used to withdraw the amniotic fluid. The sample is sent to a laboratory, where the fetal cells are grown for up to 3 weeks to obtain enough cells to study. Tests that can be used to study a fetus include chromosome analysis, biochemical analysis, amniotic fluid analysis, and DNA analysis.

Chromosome analysis

Chromosome analysis (see page 108) of fetal cells can usually be performed a couple of weeks after the cells are obtained. Using chromosome studies, doctors can identify all the known chromosome abnormalities, including Down's syndrome. Although chromosome disorders cannot yet be treated, prenatal testing gives couples the opportunity to make decisions about how to proceed.

Biochemical analysis

Many inherited biochemical disorders (referred to as inborn errors of metabolism), including Tay-Sachs disease, can be diagnosed before birth by measuring the level of activity of specific enzymes or other proteins in fetal cells grown in the laboratory. Biochemical analysis (see page 108) is usually carried out only if a family has a history of a particular metabolic disorder. Blood tests performed on the parents can help to determine whether one or both carry the disease-causing gene.

Amniotic fluid analysis

Amniotic fluid is sometimes studied when a fetus is thought to be at high risk of a neural tube defect. For example, if measurement of the alpha-fetoprotein level (see page 124) in a pregnant woman's blood shows it is raised, and an ultrasound scan of the fetus's head and spine fails to find a neural tube defect, the doctor will often recommend amniocentesis to determine the level of alpha-fetoprotein in the amniotic fluid. If the level in the fluid is also found to be raised, the parents are provided counseling to discuss the possibility of their having a baby with a neural tube defect or another serious abnormality.

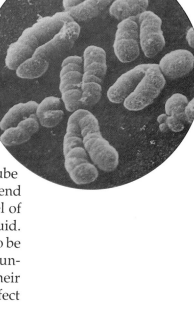

Looking at chromosomes
In the laboratory, fetal cells are examined under a powerful microscope to check for abnormalities in the number and structure of chromosomes.

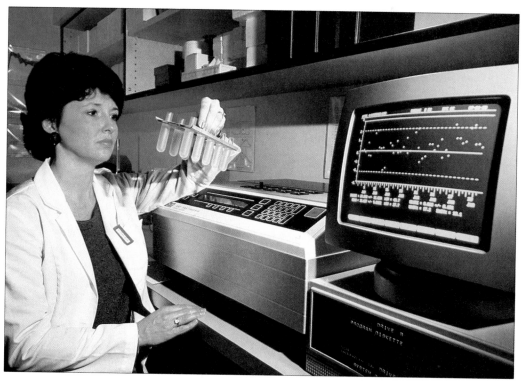

Measuring protein levels
Today, most chemical analyses of the proteins in tissue samples from a woman or fetus are performed by machines that can detect tiny amounts of the specific proteins (such as hormones or enzymes) under study.

DNA analysis

DNA analysis involves the use of special enzymes and genetic probes (see page 109) to identify people who have a particular gene defect. Thanks to continuing improvements in this technique, an increasing number of single-gene disorders can be diagnosed prenatally. Scientists have identified many disease-causing genes, as well as the specific defects in those genes.

CHORIONIC VILLUS SAMPLING

Chorionic villus sampling is a relatively new technique that can be performed at an earlier stage of pregnancy than amniocentesis – as early as the ninth week. A small sample of placental tissue, consisting of threadlike projections called chorionic villi, is removed from the uterus in one of two ways. In transcervical chorionic villus sampling, which is the technique used most often, a thin tube is passed up through the vagina and cervix. In transabdominal chorionic villus sampling, a hollow needle is inserted through the abdomen, as in amniocentesis. During both procedures, an ultrasound scanner (see page 125) monitors the position of the tube or needle.

Cells from the sample are examined in the laboratory. Chromosome analysis, DNA analysis, and biochemical studies may be performed on the tissue sample, as in amniocentesis. The cells of the placental tissue contain the same genetic material as the cells of the fetus. The main risk associated with chorionic villus sampling is that an average of two women in 100 miscarry – a risk four times greater than that associated with amniocentesis. However, some couples are willing to take the risk because chorionic villus sampling can be performed earlier in pregnancy than amniocentesis and, because the tests can be performed directly on the extracted cells, some of the results are available as soon as 7 days

after the procedure. If a severe disorder is detected, a couple has the option of terminating the pregnancy at an earlier, medically safer stage for the woman.

FETAL BLOOD SAMPLING

A sample of a fetus's blood is taken when results obtained by amniocentesis or chorionic villus sampling are ambiguous or when test results are needed quickly at a later stage of pregnancy. Using ultrasound guidance, a needle is inserted through a woman's abdominal wall to extract a sample of fetal blood from the umbilical cord. The procedure can be done safely only after about 18 weeks of pregnancy. Laboratory tests that can be performed on a fetal blood sample include chromosome analysis and tests to detect Rh (rhesus factor) blood group

Chorionic villus sampling
In transcervical chorionic villus sampling, a thin tube attached to a syringe is passed up through the vagina and cervix into the uterus. The syringe is used to suck a few threadlike pieces of tissue from the fetal side of the placenta. The procedure is guided by ultrasound. The process causes slight discomfort similar to that experienced during a Pap smear.

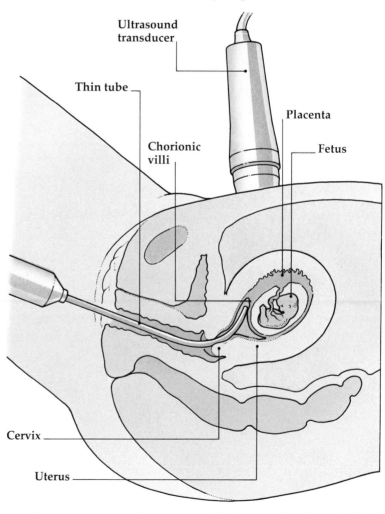

Ultrasound transducer

Thin tube

Chorionic villi

Placenta

Fetus

Cervix

Uterus

incompatibility between the fetus and the pregnant woman, the blood diseases thalassemia and sickle cell anemia, immune system disorders, and infections. The test results are usually available 48 hours after the procedure. The risk of miscarriage from fetal blood sampling is about 2 in 100.

FETOSCOPY

Fetoscopy is a highly specialized and rarely performed procedure in which the fetus is observed directly inside the uterus through a flexible viewing tube called a fetoscope. Because fetoscopy carries a higher risk of miscarriage (between 3 and 6 in 100) than other prenatal tests, it has now been almost entirely replaced by ultrasound scanning, which is an easier, safer method of observing the fetus. However, fetoscopy is used in rare cases to try to surgically correct a serious congenital defect, such as a urinary system disorder.

Fetal blood sampling
A fetal blood sample is taken only when safer procedures have failed to produce a sample that provides a definite diagnosis. The woman is given a local anesthetic and, under ultrasound guidance, a long needle is inserted through her abdomen. The blood is taken from the umbilical cord close to the point where it connects with the placenta.

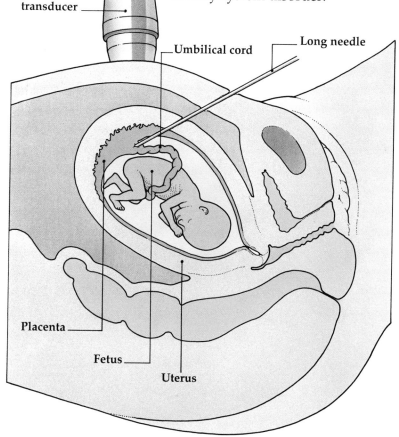

Ultrasound transducer

Umbilical cord

Long needle

Placenta

Fetus

Uterus

ASK YOUR DOCTOR
PRENATAL SCREENING

Q I have just received the results of my amniocentesis. My doctor tells me that the fetus's chromosomes are normal. Does this mean that my baby will be healthy?

A The test results show that the fetus does not have a chromosome abnormality. Ask your doctor if other tests were also performed. Negative results from amniocentesis do not guarantee a healthy baby because there are many conditions that the procedure does not or cannot test for.

Q My brother died of cystic fibrosis. If I have children, are they at risk of having it? What can I do about it if they are?

A Because the gene for cystic fibrosis has been identified, testing can establish whether or not you are a carrier. The gene is recessive, so you and your partner would both have to be carriers for your children to be at risk. You and your partner can be tested and, if you are found to be carriers, a genetic counselor can inform you of the family planning options available to you.

Q My doctor has told me that my child is at increased risk of Down's syndrome, which can be detected during pregnancy by one of two prenatal diagnostic procedures. Which method is safer?

A Amniocentesis carries a lower risk of miscarriage than chorionic villus sampling but it is usually not performed until the 16th week of pregnancy. Chorionic villus sampling can be performed as early as the ninth week. However, the procedure carries a higher risk of infection in the pregnant woman and of miscarriage.

TREATING INHERITED DISORDERS

NEW TECHNIQUES in gene therapy may someday provide cures for inherited disorders. In the meantime, doctors can alleviate the symptoms of many genetic diseases with treatments including medication, special diets, and surgery. Many metabolic disorders are treated with supplements of missing proteins.

Inherited disorders produce a wide range of symptoms. Some are mild; others are very severe. Many people, such as those with lactose intolerance, can expect to lead normal lives with a minimum of medical care. Other people may be seriously affected by a genetic disorder and may require lifelong medical treatment.

CONVENTIONAL TREATMENTS

Most conventional treatments for genetic disorders are designed to relieve the symptoms of the disease, to prevent complications from developing, to improve and prolong the life of the individual, and to lessen the burden of the disorder on a family. A combination of treatments is often used, depending on the disorder. For example, with the help of a special diet, physical therapy, and antibiotics to prevent lung infections, people with cystic fibrosis now have a much improved quality of life and they live longer. Surgery is used to correct some life-threatening congenital abnormalities such as pyloric stenosis, an irregularity in the opening between the stomach and intestines that prevents food from leaving the stomach.

INBORN ERRORS OF METABOLISM

Inborn errors of metabolism are imbalances in the body's chemistry caused by a defective gene. Such biochemical imbalances often result from a deficient or inefficient enzyme that controls a specific chemical reaction inside the body. Some inborn errors of metabolism can now be effectively treated with a restricted diet or supplements of a missing enzyme.

Manipulating enzymes

In some inborn errors of metabolism, the normal quantity of an enzyme is produced, but its action is less efficient than usual. An abnormal enzyme can sometimes be made more effective by supplying an activator such as a vitamin. For example, homocystinuria (a disorder associated with skeletal deformities and mental retardation) is caused by an

Adapting to special needs
By adapting the home environment to their particular needs, many people with physical disabilities can be more independent. In a similar way, special education programs help people with mental handicaps do more for themselves.

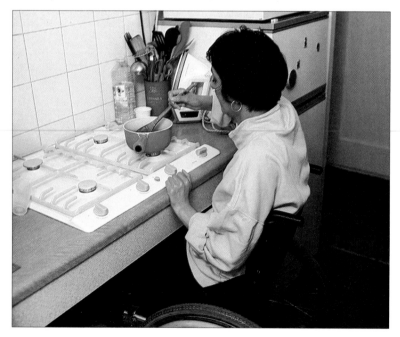

enzyme deficiency that leads to a buildup of an amino acid called homocystine. In some people, the symptoms can be relieved with vitamin B_6 supplements.

Preventing buildup of poisons

Restricted diets are used to treat many metabolic disorders, including phenylketonuria (PKU). In PKU, an enzyme deficiency causes the amino acid phenylalanine to build up in the blood and nervous system, which leads to mental retardation. But when the disorder is identified at birth and the child is put on a diet that excludes the amino acid, he or she usually develops normal or almost normal intelligence. In the US, all newborns are screened for PKU.

Drugs are sometimes used to promote the breakdown or elimination of a poisonous substance from the body. In Wilson's disease, a buildup of copper inside the body leads to brain and liver damage. The drug penicillamine, which a person must take for life, prevents the harmful accumulation by eliminating copper from those organs.

TREATABLE DISORDERS

EXAMPLES	TREATMENT
Adult polycystic disease; cystinosis	Kidney transplant
Beta-thalassemia	Bone marrow transplant, if necessary
Cleft lip and palate	Surgery
Congenital adrenal hyperplasia	Hormone treatment
Cystinuria	Penicillamine and extra fluids
Diabetes; familial hypercholesterolemia	Diet and medication
Epilepsy	Medication
Gaucher's disease	Replacement of missing enzyme
Hemophilia	Replacement of factor VIII (the missing blood-clotting protein)
Hereditary spherocytosis	Spleen removal, if necessary
Homocystinuria	Vitamin B_6 supplements
Phenylketonuria; galactosemia	Diet restrictions
Wilson's disease	Penicillamine
Xeroderma pigmentosum	Avoiding exposure to the sun

ENZYMES: CONTROLLERS OF CHEMICAL REACTIONS

Enzymes are specialized proteins that control chemical reactions inside the body. Each enzyme-controlled reaction is one of a chain of reactions that converts an initial substance into an end product. A single defective gene may not make enough of a particular enzyme or may make one that is defective. The resulting enzyme deficiency causes a chemical imbalance called an inborn error of metabolism.

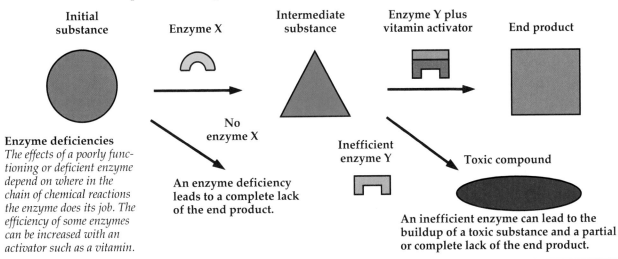

Initial substance   **Enzyme X**   **Intermediate substance**   **Enzyme Y plus vitamin activator**   **End product**

No enzyme X

Enzyme deficiencies
The effects of a poorly functioning or deficient enzyme depend on where in the chain of chemical reactions the enzyme does its job. The efficiency of some enzymes can be increased with an activator such as a vitamin.

An enzyme deficiency leads to a complete lack of the end product.

Inefficient enzyme Y

Toxic compound

An inefficient enzyme can lead to the buildup of a toxic substance and a partial or complete lack of the end product.

REPLACING CHEMICAL DEFICIENCIES

Some metabolic disorders lead to the deficiency of a particular protein. These disorders can often be treated by replacing the missing protein, either by mouth or by injection. Some children are born with a deficiency of thyroid hormone because an enzyme required for making the hormone is absent. Taking oral doses of the hormone can prevent the growth and mental retardation that these children would otherwise have.

Hemophilia, a serious blood disorder, can now be treated by giving people an intravenous supply of factor VIII, the blood-clotting protein that they lack.

Enzyme replacement

Direct replacement of a missing enzyme is proving successful in treating a few metabolic disorders, including the adult form of Gaucher's disease.

Porphyria
Porphyria is a disorder associated with the accumulation in the body of substances called porphyrins. Because these substances react with sunlight, people with the disorder develop rashes and blisters on their skin when they are exposed to sunlight. Avoiding the sun can prevent the symptoms.

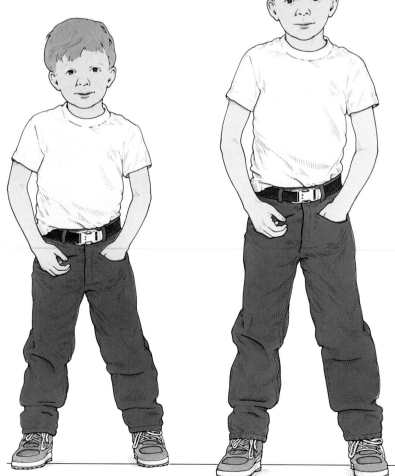

Growth hormone
If the growth of a child is retarded by a deficiency of growth hormone, that child can achieve almost normal height when given intramuscular injections of synthetically produced growth hormone.

Untreated Gaucher's disease can cause enlargement of the spleen and liver, anemia, and bone damage.

However, in disorders associated with nerve damage in the brain, enzyme replacement is not effective. Supplementary enzymes absorbed into the blood cannot always enter the brain because of a special protective barrier in the small blood vessels of the brain.

Sources of replacement proteins

In the past, some of the proteins required for replacement therapy were obtained from cadavers or from other human sources, such as donor blood. However, the use of such products carries the risk of transmitting infections. This danger was demonstrated recently when a number of people being treated with growth hormone from cadavers were accidentally infected with the organism that causes a brain-destroying illness. The use of growth hormone from human sources was immediately stopped.

Advances in genetic engineering are enabling scientists to produce an increasing number of proteins in the laboratory. Genetically engineered proteins can be used safely in replacement therapies. For example, a synthetic form of human growth hormone is being used to treat children whose growth would otherwise be stunted because their bodies do not produce enough of the hormone. A genetically engineered form of the blood-clotting protein factor VIII is now used to treat people with hemophilia.

ORGAN TRANSPLANTS

When an organ is damaged by a genetic disorder, it is sometimes possible for doctors to remove it and replace it. For example, a kidney transplant can be life-saving for people with an inherited disorder such as polycystic kidney disease, which eventually leads to severe kidney damage. However, all organ transplants carry the risk that the recipient's immune system will reject the transplanted organ. To reduce the risk of organ rejection, recipients are given powerful drugs that suppress their immune system.

Liver transplant

Much of the body's most important biochemical processing takes place in the liver. If a certain protein, such as an enzyme, is lacking, many of these processes can be disrupted. For some inherited metabolic disorders that damage the liver, such as Wilson's disease and alpha$_1$-antitrypsin deficiency, a liver transplant may be the only effective treatment. However, liver transplantation has a high failure rate – 20 percent of people require a second transplant and between 10 and 20 percent die.

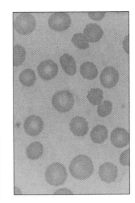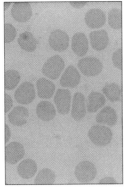

Bone marrow transplant

Healthy bone marrow produces the white blood cells that fight infection, the oxygen-carrying red blood cells, and platelets to help blood clot. In blood disorders such as thalassemia, sickle cell anemia, and osteopetrosis (a bone marrow disorder), the normal production of blood cells is disrupted. To provide adequate blood cell formation, a bone marrow transplant is sometimes necessary.

The main problem associated with bone marrow transplants is finding a donor whose cells and tissues match those of the recipient. If the tissues do not match, the recipient's immune system is likely to reject the donated bone marrow. When a matched donor cannot be found in a potential recipient's family, transplant experts explore the possibility of using compatible but unrelated donors.

Using bone marrow from an unrelated donor increases the risk associated with this type of transplant, which is fatal in one out of five cases. The risks of undergoing the procedure need to be carefully weighed against the likely course of the disease if it is not treated.

Improved transplant techniques
Improvements in surgical techniques have made it possible to divide an adult liver into sections that can be implanted into different children. In this way, one donated adult liver can provide transplants for more than one child.

Hereditary spherocytosis
In hereditary spherocytosis, the red blood cells are spherical (far left).These very fragile cells are easily destroyed in the spleen, causing anemia. Normal red blood cells (near left) have a flat, doughnut shape that enables them to pass through tissues.

FINDING A DONOR

The more siblings a person has, the greater the chance of finding a compatible donor for a bone marrow transplant. Parents are rarely suitable donors because they share only half (at most) of the factors that determine compatibility.

A person with three siblings has about a 60 percent chance that at least one will be a compatible donor.

With one sibling, there is a 25 percent chance that he or she will be a compatible donor.

CASE HISTORY
BONE MARROW TRANSPLANT

FROM A YOUNG AGE, **Costas regularly received blood transfusions to treat thalassemia, a hereditary disorder of red blood cells. But lately, even right after having a transfusion, he would complain about feeling weak and tired. Also, his parents noticed that his skin was grayish tan and the whites of his eyes were a muddy yellow.**

PERSONAL DETAILS
Name Costas Dimitrios
Age 6
Occupation Student
Family His uncle died of thalassemia at a young age.

MEDICAL BACKGROUND
By the time he was 9 months old, Costas had already received seven blood transfusions for severe anemia. Tests confirmed that the anemia was caused by the genetic blood disorder thalassemia. Thalassemia prevents the body's bone marrow from making an adequate supply of hemoglobin, the oxygen-carrying substance in red blood cells.

AT THE DOCTOR'S OFFICE
The doctor tells Costas' parents that his repeated blood transfusions have led to a buildup of iron in his body. The doctor recommends that Costas have a bone marrow transplant as an alternative treatment and he explains the risks and benefits of the procedure. The couple feels that the potential improvement in their son's quality of life far outweighs the risks, and they decide to give permission for him to have a bone marrow transplant as soon as a compatible donor can be found.

FURTHER INVESTIGATIONS
Tests show that one of Costas' sisters, Helen, can provide compatible bone marrow for his transplant.

THE TRANSPLANT
Two weeks before the transplant, Costas is admitted to the hospital. While he is under anesthesia, a tube is inserted into a large vein near his heart. The tube allows the doctors to give Costas intravenous medication and an intravenous diet, and to periodically remove samples of blood. During this time, Costas is being treated with powerful drugs and radiation that destroy his defective bone marrow. Because his immune system stops functioning when the bone marrow is destroyed, Costas is kept in a sterile isolation room to protect him from germs.

Costas' sister Helen is given a general anesthetic. Using a syringe, doctors withdraw a small amount of bone marrow from her hipbone. A solution containing her bone marrow is then given to Costas intravenously through the tube in his chest. He is also given a drug to prevent his immune system from rejecting the donated bone marrow cells.

THE OUTCOME
After 10 days, samples of Costas' blood show that his body has accepted his sister's cells. By the 15th day, his white blood cell count is rising steadily – an indication that his bone marrow and his immune system are functioning normally. He is able to leave the hospital 3 weeks after the transplant. A year later, Costas is healthy and active and does not need to have any more blood transfusions.

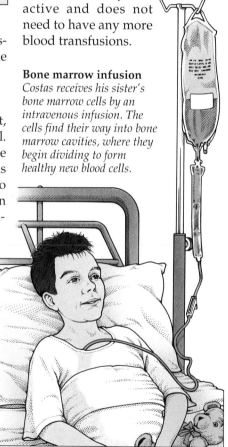

Bone marrow infusion
Costas receives his sister's bone marrow cells by an intravenous infusion. The cells find their way into bone marrow cavities, where they begin dividing to form healthy new blood cells.

GENE THERAPY

Although only a theoretical glimmer right now, it may one day be possible to prevent genetic disorders by identifying disease-causing genes in embryos and replacing them with healthy genes.

Transplanting genes

One form of gene therapy currently under investigation is similar to an organ transplant. In this technique, genes are introduced into the affected body cells, such as bone marrow cells in the blood disorder thalassemia. A virus that has been made harmless can be used to transport the gene into the cells' DNA. However, without adequate safeguards, inserting a gene into a person's cells might result in replacing one genetic disorder with another, or even causing cancer. Furthermore, because only the cells in specific tissues are treated, a person's sperm or egg cells would still carry the defective gene. Although the gene would no longer affect that person, it could be passed on to children who would carry the defect in all of their cells.

How gene therapy works
One way in which gene therapy might work is shown below. Using a virus as the means of transportation, a healthy version of the defective gene is inserted into a sample of cells from the person's bone marrow. The altered bone marrow cells would then be reintroduced into the person. The normal gene in these cells, or in the blood cells derived from them, would produce the protein that the defective gene could not.

Person undergoing gene therapy.

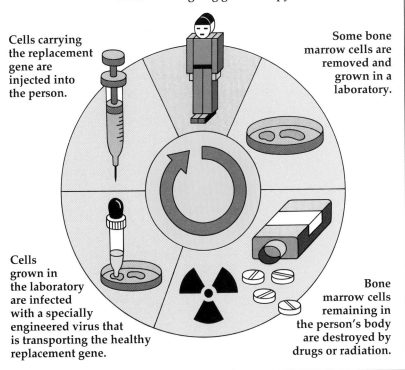

Cells carrying the replacement gene are injected into the person.

Some bone marrow cells are removed and grown in a laboratory.

Cells grown in the laboratory are infected with a specially engineered virus that is transporting the healthy replacement gene.

Bone marrow cells remaining in the person's body are destroyed by drugs or radiation.

Q **My son has hemophilia. Could he get AIDS from his factor VIII replacement therapy?**

A It's highly unlikely. Replacement factor VIII comes from pooled donated blood that has been screened for the presence of disease organisms. The blood product is purified to avoid the risk of contamination with the AIDS virus. Factor VIII is now also being produced by genetic engineering; this product carries no risk of contamination at all.

Q **My son has galactosemia and is on a milk-free diet. Why can't he be given the missing enzyme as a medicine instead?**

A Your son cannot take the enzyme by mouth because acid in the stomach would break it down. Nor can he be given the enzyme by injection because the liver would trap it and break most of it down before it had a chance to work. Your son's restricted diet makes it less critical that he lacks the specific enzyme because the diet prevents the serious complications (liver disease, cataracts, and mental retardation) that can result from galactosemia.

Q **My baby daughter was diagnosed as having congenital adrenal hyperplasia. Can you tell me what will happen to her?**

A In severe cases, hormone imbalances associated with congenital adrenal hyperplasia cause masculinization of the genitals. However, because the disorder was diagnosed early, your daughter can lead a normal, healthy life with lifelong hormone-replacement therapy.

FACT FILE OF SINGLE-GENE DISORDERS

This section contains information about some of the more than 4,000 genetic disorders caused by single harmful genes. Many of the most common single-gene disorders have been discussed elsewhere in this volume. In addition to the patterns of inheritance described in the box below, single-gene disorders can also result from new mutations that occur during egg or sperm formation in an unaffected parent.

PATTERNS OF INHERITANCE

Autosomal recessive – The disease gene is carried on a nonsex chromosome (autosome) and inherited from two usually unaffected parents.

Autosomal dominant – The disease gene is carried on a nonsex chromosome (autosome) and inherited from one usually affected parent.

X-linked recessive – The disease gene is carried on the X sex chromosome and inherited by a son from a usually unaffected mother.

ADRENOLEUKODYSTROPHY

Inheritance: X-linked recessive.
Features: Brain disturbances that become progressively more severe, sometimes accompanied by symptoms such as muscle weakness resulting from reduced hormone production by the adrenal glands.
Diagnosis and treatment: Diagnosed from levels of specific fatty acids in the blood. Carriers can be detected by both DNA analysis and biochemical analysis. Prenatal diagnosis is possible by first determining the sex of the fetus. If the fetus is male, fetal cells can be analyzed biochemically. Hormone replacement therapy can help with the symptoms caused by the hormone deficiency, but there is no treatment for the nervous system symptoms.

AGAMMAGLOBULINEMIA

Inheritance: X-linked recessive.
Features: The immune system fails to make antibodies that fight bacterial infections.
Diagnosis and treatment: Prenatal diagnosis is possible using fetal blood sampling. Treatment involves regular injections of antibodies to fight common bacterial infections.

ALBINISM

Inheritance: Autosomal recessive.
Features: Absence of pigment in skin and hair and the irises of the eyes; vision problems that can affect reading; susceptibility to skin cancer. Albinism affects all ethnic groups.
Diagnosis and treatment: Carrier detection is not possible. Vision problems can be alleviated with glasses. Guarding against sun exposure with protective clothing and sunscreen reduces skin cancer risk.

ALPHA₁-ANTITRYPSIN DEFICIENCY

Inheritance: Autosomal recessive.
Features: Liver disease and a susceptibility to the lung disease emphysema between ages 30 and 50.
Diagnosis and treatment: Prenatal diagnosis is possible using DNA analysis. Treated by not smoking, enzyme replacement therapy (to delay the onset of emphysema), and liver transplantation.

ALPHA-THALASSEMIA

Inheritance: Autosomal recessive.
Features: Caused by a deficiency or lack of a type of hemoglobin (the oxygen-carrying pigment in red blood cells). The most severe form causes death in an affected fetus or death soon after birth. A less severe form causes a type of anemia.
Diagnosis and treatment: Carrier detection is not always possible but prenatal diagnosis of the most severe form is possible.

ALPORT'S SYNDROME

Inheritance: Autosomal dominant; X-linked recessive.
Features: Abnormal production of a protein essential to membrane formation in the kidneys and the inner ear. Affected people have kidney disorders and hearing loss.
Diagnosis and treatment: Carriers can be detected by DNA analysis. Diagnosis is confirmed by a kidney biopsy. There is no treatment other than kidney transplantation.

CONGENITAL ADRENAL HYPERPLASIA

Inheritance: Autosomal recessive.
Features: Abnormal production of some steroid hormones by the adrenal glands, which causes vomiting and shock in young babies, enlarged genitals in males, and masculinization of the genitals in females.

Diagnosis and treatment: Carriers can be detected using DNA analysis; prenatal diagnosis is possible using a combination of DNA markers (see page 112) and biochemical analysis. Treatment involves lifelong replacement of the missing hormones.

CYSTINURIA

Inheritance: Autosomal recessive.
Features: Recurrent stones in the kidneys and bladder, caused by the inability of the kidneys to absorb the essential amino acid cystine.
Diagnosis and treatment: Carrier detection and prenatal diagnosis are possible using biochemical analysis. Treatment involves drinking plenty of fluids and taking medication to prevent the formation of stones.

EHLERS-DANLOS SYNDROME

Inheritance: Usually autosomal dominant.
Features: There are several forms of the disorder, each of which may cause some or all of the following symptoms – elasticity of the skin; loose joints and multiple joint dislocations; fragile, easily bruised skin; and slow healing of wounds.
Diagnosis and treatment: Carrier detection using DNA analysis is possible for some forms. There is presently no treatment.

FAMILIAL HYPERCHOLESTEROLEMIA

Inheritance: Autosomal dominant.
Features: Increased levels of fat in the blood, causing risk of early death from coronary heart disease.
Diagnosis and treatment: Prenatal diagnosis is possible using DNA analysis; diagnosis may be made at birth by measuring cholesterol levels in blood samples from the umbilical cord or by measuring blood cholesterol levels in children and adults. Treatment involves eating a diet low in cholesterol and saturated fats and taking cholesterol-lowering drugs.

FRIEDREICH'S ATAXIA

Inheritance: Autosomal recessive.
Features: Degeneration of nerve fibers in the spinal cord causes loss of balance and coordination, slurred speech, and rapid, involuntary eye movements. The disorder can lead to heart defects and diabetes.
Diagnosis and treatment: Carrier detection and prenatal diagnosis are possible using DNA analysis. There is presently no treatment.

GALACTOSEMIA

Inheritance: Autosomal recessive.
Features: The lack of a specific enzyme prevents the body from breaking down the milk sugar galactose. Accumulation of galactose, which is harmful to many tissues, leads to vomiting and diarrhea and stunts growth. Liver damage, cataracts, and mental retardation result. If not treated until after the first month of life, the damage is permanent.
Diagnosis and treatment: Prenatal diagnosis is possible using biochemical analysis. The disorder is treated with a milk-free diet. A pregnant woman who is a carrier and has an affected fetus should eliminate milk from her diet.

GAUCHER'S DISEASE

Inheritance: Autosomal recessive.
Features: Enlargement of the liver and spleen, bone pain, and anemia. Some forms affect the nervous system, causing symptoms such as progressive dementia, eye movement disorders, and epilepsy.
Diagnosis and treatment: Carrier detection and prenatal diagnosis are possible using biochemical analysis or DNA analysis. Replacement of the missing enzyme helps treat some forms of the disease.

HEMOCHROMATOSIS

Inheritance: Autosomal recessive.
Features: The disorder affects males mostly. Excess iron is stored in many organs – including the liver, pancreas, heart, and pituitary gland – causing serious tissue damage; results in cirrhosis of the liver, diabetes, heart failure, and arthritis.
Diagnosis and treatment: Detection of carriers and diagnosis is made using biochemical analysis; treatment involves periodic removal of a specified amount of blood to remove excess iron from the body.

HUNTINGTON'S CHOREA

Inheritance: Autosomal dominant.
Features: Progressive loss of nerve cells in the brain, which causes involuntary jerking of the limbs and facial contortions, speech defects, and irregular breathing. Symptoms usually appear between ages 35 and 50; they become increasingly severe and are associated with dementia. Death usually occurs within 15 to 20 years of first symptoms.
Diagnosis and treatment: DNA analysis is used for prenatal diagnosis and to determine if a person has the gene and, therefore, will eventually develop the disorder. There is presently no treatment.

HYPERTROPHIC CARDIOMYOPATHY

Inheritance: Autosomal dominant.
Features: Excessive growth of muscle fibers causes areas of the heart to thicken but seldom produces symptoms. However, the disorder usually causes sudden death. If symptoms do occur, the most common are breathlessness, chest pain, fatigue, and fainting.
Diagnosis and treatment: Diagnosis can be confirmed by ultrasound scanning of the heart. Heart transplantation is a possible treatment.

MALIGNANT HYPERTHERMIA

Inheritance: Autosomal dominant.
Features: Symptoms develop only when an affected person is given the anesthetic drugs halothane or succinylcholine during general anesthesia. Powerful contractions occur in the muscles, resulting in excessive heat production. The person's body temperature rises to a dangerous level that, unless treated immediately, can be fatal.
Diagnosis and treatment: DNA analysis and muscle biopsy are used for diagnosis.

MARFAN'S SYNDROME

Inheritance: Autosomal dominant.
Features: Long limbs and fingers, loose joints, dislocation of the lenses of the eyes, and weakness of the heart valves and aorta.
Diagnosis and treatment: Prenatal diagnosis may be possible using DNA analysis. Drugs can be taken for heart symptoms.

MUCO-POLYSACCHARIDOSIS

Inheritance: Autosomal recessive; type II is X-linked recessive.
Features: There are five main forms of mucopolysaccharidosis; type I (called Hurler's syndrome) is the most recognizable and most serious. All forms are caused by deficiencies of related enzymes required to break down an important component of connective tissue. Each form has its own features, but most include mental retardation, bone deformities, disorders of the heart and large arteries, and clouding of the corneas. Many children have stunted growth and abnormal facial features. Their liver and spleen are usually enlarged.
Diagnosis and treatment: Prenatal diagnosis is possible using amniotic fluid analysis or biochemical analysis. Carrier detection is possible using biochemical analysis or DNA analysis. There are no treatments for any of the forms.

MULTIPLE ENDOCRINE NEOPLASIA

Inheritance: Autosomal dominant.
Features: Tumors form in hormone-secreting glands such as the pituitary, pancreas, thyroid, parathyroid, or adrenal glands. The outlook varies, depending on whether the tumors are cancerous.
Diagnosis and treatment: Prenatal diagnosis and diagnosis before a person has symptoms are possible using DNA analysis or biochemical analysis. Treatment includes removing tumors and replacing missing hormones when necessary.

MYOTONIC DYSTROPHY

Inheritance: Autosomal dominant. Risk varies depending on whether the defective gene is inherited from the mother or from the father.
Features: Progressive weakness of muscles in the face, neck, arms, and hands; cataracts; difficulty relaxing the grip of the hands. The heart and the retinas of the eyes can also be affected. General anesthesia can be life-threatening. Most people become severely disabled within 20 years of the first symptoms.
Diagnosis and treatment: Detection of carriers and prenatal diagnosis are possible in some families using DNA analysis.

NOONAN'S SYNDROME

Inheritance: Sometimes autosomal dominant but often no clear pattern of inheritance.
Features: Affected people have short stature, drooping of the upper eyelid, webbed neck, heart abnormalities, and, in some cases, mild mental retardation. Affected males often have undescended testicles. **Diagnosis and treatment**: Diagnosis is made by physical examination. There is presently no treatment.

OSTEOGENESIS IMPERFECTA

Inheritance: Autosomal dominant in most cases.
Features: The bone disorder results from a defect in the formation of the connective tissue collagen, causing bones to break easily. The whites of the person's eyes are so thin that the underlying pigment shows through, producing a blue appearance. One form (type II) is much more severe; most affected children are stillborn or die shortly after birth.
Diagnosis and treatment: Carrier detection and prenatal diagnosis are possible using biochemical analysis or DNA analysis. Prenatal diagnosis of type II is also possible with ultrasound. Treatment involves correcting fractures surgically.

OTOSCLEROSIS

Inheritance: Autosomal dominant.
Features: Progressive hearing loss in middle age caused by fusion of bones in the middle ear.
Diagnosis and treatment: Diagnosis can be made by physical examination and hearing tests. Treatment involves using a hearing aid or surgically replacing abnormal bones.

PERONEAL MUSCULAR ATROPHY

Inheritance: Autosomal dominant; occasionally autosomal recessive or X-linked recessive.
Features: Also called Charcot-Marie-Tooth disease, the disorder results from degeneration of some of the peripheral nerves. It causes muscle wasting, first in the feet and calves and then in the hands and arms.
Diagnosis and treatment: Diagnosis can be confirmed by nerve biopsy. There is presently no treatment.

RETINITIS PIGMENTOSA

Inheritance: Autosomal dominant; autosomal recessive (70 percent); X-linked recessive.
Features: Progressive loss of vision.
Diagnosis and treatment: The pattern of inheritance can often be determined by studying the family history. For the X-linked form, DNA analysis can be used, together with an eye examination, to identify female carriers. DNA analysis can also be used in some families with the autosomal dominant form.

RETT SYNDROME

Inheritance: Thought to be X-linked dominant; the defective gene is lethal in male embryos.
Features: Brain dysfunction first appears between the ages of 6 months and 4 years, when affected girls stop developing normally. Usually, by age 25, they have severe mental and physical disability.
Diagnosis and treatment: Difficult to detect in tests. Seizures, beginning between ages 4 and 7, can be controlled with medication. There is no treatment to halt the progress of the disease.

SEVERE COMBINED IMMUNODEFICIENCY

Inheritance: Autosomal recessive; some forms are X-linked recessive.
Features: Severe inability to develop immunity caused by abnormal formation of some types of white blood cells. After the first few months of life, an affected baby becomes susceptible to infections.
Diagnosis and treatment: Carriers of X-linked forms can be identified using DNA analysis. Prenatal diagnosis of some autosomal recessive forms is possible using biochemical analysis of amniotic fluid or fetal blood sampling. The only effective treatment is a bone marrow transplant.

SICKLE CELL ANEMIA

Inheritance: Autosomal recessive.
Features: Occurs primarily in blacks. The disorder reduces the amount of oxygen that can be carried in red blood cells, which makes them become distorted. The abnormal sickle shape prevents the cells from passing through small blood vessels and shortens their normal 4-month life span to a few weeks. The result is severe anemia, blockage of blood flow to various organs, and damage to tissues. People who carry only one copy of the defective gene are said to have sickle cell trait, which seldom causes symptoms.
Diagnosis and treatment: Prenatal diagnosis is possible; carrier status can be detected with a simple blood test. Such measures as drugs, oxygen therapy, and blood transfusions may be required for long-term treatment or to treat an obstruction in blood flow whenever it occurs.

SPINAL MUSCULAR ATROPHY

Inheritance: Autosomal recessive; there is an X-linked recessive form.
Features: Cells that form the motor nerves in the spinal cord (which control muscle movement) are gradually destroyed, resulting in progressive muscle wasting. Facial tremors are common. Intelligence is not affected. In the more severe form, infants die of respiratory failure.
Diagnosis and treatment: Carrier detection and prenatal diagnosis are possible in some families using DNA analysis. There is presently no specific treatment.

TOURETTE'S SYNDROME

Inheritance: Thought to be autosomal dominant.
Features: Recurrent, uncontrollable twitching along with snorts and sniffs and the shouting of obsceni-

ties; obsessive-compulsive behavior; learning difficulties.
Diagnosis and treatment: Difficult to detect in tests. Different drugs have proved effective to varying degrees.

TREACHER COLLINS SYNDROME

Inheritance: Autosomal dominant; most cases are new mutations.
Features: Severe skull deformities that include flattened cheekbones, a small jaw, eyes that slope downward, notches in the lower lids, and ear malformations. A third of affected people have cleft palate and a third are deaf. About 5 percent have mental retardation.
Diagnosis and treatment: Because most cases are new mutations, they are difficult to detect. There is no specific treatment but facial malformations may be corrected by surgery.

TUBEROUS SCLEROSIS

Inheritance: Autosomal dominant; most cases are new mutations.
Features: Symptoms vary in severity. They include white skin patches, facial rash, swellings (particularly around the nose and chin), mental retardation, epilepsy, and tumors in the brain and kidneys.
Diagnosis and treatment: Diagnosis is made by physical examination and confirmed by imaging of the brain with computed tomography (CT) or magnetic resonance imaging (MRI).

VON WILLEBRAND'S DISEASE

Inheritance: Autosomal dominant; a severe form is autosomal recessive.
Features: Insufficient production of a protein required for blood clotting, resulting in excess bleeding after injury or surgery.
Diagnosis and treatment: Diagnosis is possible using biochemical analysis. Treatment involves infusions of the deficient protein.

GLOSSARY OF TERMS

Terms in *italics* refer to other terms in this glossary.

A

Alpha-fetoprotein
A *protein* produced by a fetus and found in amniotic fluid. Measuring the level in a pregnant woman's blood can help detect some abnormalities in a fetus.

Amino acids
The building blocks of *proteins*. The 20 common amino acids string together in specified patterns to form different proteins.

Amniocentesis
A prenatal diagnostic procedure, usually performed during the 16th week of pregnancy, in which amniotic fluid is withdrawn from the amniotic sac surrounding the fetus.

Amplification
The rapid production of millions of copies of a piece of *DNA*.

Autosomes
All the *chromosomes* in the human cell other than the *sex chromosomes* X and Y.

B

Base triplet
A group of three *nucleotide bases* found in *DNA* and *RNA* that is the code for a specific *amino acid* in *protein* production.

Bases
See *nucleotide bases*.

Birth defect
Any abnormality that is present at birth; a *congenital abnormality*. A birth defect may or may not be inherited and may or may not be obvious at birth.

C

Carrier
A person who possesses, or carries, a *gene* for a particular disorder but is not affected by the disorder.

Chorionic villus sampling
A prenatal diagnostic procedure, performed in the ninth to 11th week of pregnancy, in which fetal tissue is withdrawn from the placenta and studied for genetic abnormalities.

Chromosomes
The structures inside the nucleus of a cell that contain the *genes*. Body cells have 46 chromosomes; egg and sperm cells have 23.

Concordance
The presence of the same trait or disorder in a pair of twins.

Congenital abnormality
An abnormality that is present at birth; a *birth defect*. A congenital abnormality may or may not be inherited and may or may not be obvious at birth.

Crossing over
Exchange of genetic material between matching pairs of *chromosomes* during egg and sperm formation.

D

Deletion
Loss of a portion of a *chromosome* or a segment of *DNA*.

DNA (deoxyribonucleic acid)
The molecular structure inside every cell, which contains all the information needed to make a person. *Genes* are made of DNA.

DNA analysis
Study of the structure of a person's *DNA*, including his or her *genetic code*.

DNA marker
A piece of *DNA* located at a particular site on a *chromosome* that varies detectably from person to person. DNA markers are used as reference points for locating disease-causing *genes*.

Dominant
Describes a *gene* that produces an effect whenever it is present; a trait or disorder caused by a gene of this type.

E

Embryo
The early, developing, fertilized egg.

Enzymes
The *proteins* that control chemical reactions in the body.

F

Familial
Occurring more often in a family than would be expected by chance; a familial disorder may be caused by *heredity* and/or environment.

Fertilization
The union of an egg and a sperm to produce the first cell of a new person (a fertilized egg).

G

Gamete
A male or female reproductive cell (sperm or egg); a gamete contains only half the number of *chromosomes* of other cells.

Gene
The portion of a *DNA* molecule that is the basic, functional unit of *heredity*; contains the instructions to make a specific *protein*.

Gene mapping
Locating *genes* on *chromosomes*.

Genetic code
The set of instructions that directs the development and functioning of a person; the universal key by which genetic information is recorded and translated in all life-forms.

Genetic engineering
A technique to manipulate a cell's genetic information either by adding new *genes* or altering genes already present. The technique enables cells to make large quantities of *proteins* that they otherwise could not produce.

Genetic probe
A labeled fragment of *DNA* that is used to detect the presence of a matching piece of DNA in a sample of a person's DNA.

Genome
The complete set of *genes* that makes up the master blueprint for an organism.

Genotype
The entire genetic makeup of a person.

Germ cell
A sperm or egg cell or a cell that divides to form a sperm or egg.

H

Heredity
The genetic transmission of traits from parents to children.

Heritability
A measure of the degree to which a trait or disorder is determined by *genes*.

Heterozygote
A person who has two different forms of the same *gene*.

Homologous chromosomes
A matching pair of *chromosomes*, one from each parent.

Homozygote
A person who has two identical forms of a *gene*.

I

Inborn error of metabolism
An inherited disorder in which a specific *enzyme* defect blocks a metabolic (biochemical) process, causing illness.

Inversion
A *chromosome* defect in which a piece of the chromosome is turned upside down, reversing the order of its genetic instructions.

K

Karyotype
A standardized, structural classification of a person's *chromosomes* in numerical order.

L

Linkage
The tendency of two *genes* that are close to each other on the same *chromosome* to be inherited together. Linked genes are used to study the transmission of a disorder in a specific family.

M

Meiosis
A type of cell division that produces egg or sperm cells, which each contain half the amount of genetic material of the original cell.

Messenger RNA
An *RNA* molecule that conveys *DNA*'s genetic instructions out of a cell's nucleus to be read and translated into chains of *amino acids* to make *proteins*.

Mitosis
The way in which body cells divide to provide for growth and development; the genetic material inside the nucleus of a cell is exactly duplicated into two identical daughter cells.

Mosaicism
The presence of two or more genetically different groups of cells in one person, all derived from the same fertilized egg.

Multifactorial inheritance
A pattern of inheritance, seen in many *birth defects* and common diseases, determined by the interaction of many *genes* and environmental factors.

Mutagen
A chemical or environmental agent that can cause genetic *mutations*.

Mutation
A change in a *gene*.

N

Nondisjunction
The failure of *homologous chromosomes* to separate (disjoin) and go into different sex cells during *meiosis*, resulting in an egg or sperm with too few or too many *chromosomes*.

Nucleotide bases
Chemicals that form the building blocks of *DNA* and *RNA*; their sequence spells out the *genetic code*.

O

Oncogene
An altered *gene* that can lead to the unregulated cell division characteristic of cancer.

P

Polymerase chain reaction
A technique to make millions of copies of a piece of *DNA* in a short time; a powerful tool that can be used to study genetic material, including disease-causing *genes*.

Proteins
Large, complex chemicals that have many roles in the structure and functioning of cells.

Proto-oncogenes
Normal *genes* that regulate cell division. If altered, they can give rise to cancer-causing *oncogenes*.

R

Recessive
Describes a *gene* that produces an effect only when it is not overridden by a corresponding *dominant gene*; a trait or disorder caused by a gene of this type. Usually, two copies of a recessive gene must be present to have an effect.

Recombinant DNA
Artificially produced *DNA*, formed by inserting a *gene* or part of a gene from one organism into the genetic material of another.

Restriction enzyme
An *enzyme* used in *DNA analysis* to cut DNA at specific points.

RNA (ribonucleic acid)
A type of genetic material that carries out the instructions of the cell's *DNA*.

S

Sex chromosomes
The X and Y *chromosomes*, which determine sex. Females have two X chromosomes; males have an X and a Y.

Sex-limited
Describes a trait or disorder that affects only one sex.

Sex-linked
Determined by a *gene* on a sex *chromosome* (usually the X chromosome).

Somatic cells
All the cells of the body other than the *germ cells*.

T

Transcription
The process by which the *genetic code* on strands of *DNA* is converted into a matching code on strands of *messenger RNA*.

Transfer RNA
An *RNA* molecule that picks up a specific *amino acid* and lines it up with other amino acids in the correct order to make a specific *protein*.

Translation
Conversion of a sequence of *amino acids* into a *protein*.

Translocation
The transfer of a piece of one *chromosome* to another. If no genetic material is lost, the rearrangement (called a balanced translocation) usually has no noticeable effect.

Trisomy
The presence of an extra *chromosome* in an otherwise normal cell.

X

X-linked
Determined by a *gene* on the X *sex chromosome*; also *sex-linked*.

Z

Zygote
A fertilized egg cell produced by the union of an egg and a sperm; the first cell of a new person.

141

INDEX

Page numbers in *italics* refer to illustrations and captions.

Photograph sources:
Art Directors Photo Library **13**
Bart's Medical Picture Library **118** (center right)
Biophoto Associates **47**; **97**; **133**
Denver Children's Hospital Cytogenetics Department **67**
Mary Evans Photo Library **74** (top)
Sally & Richard Greenhill **130**
Robert Harding Picture Library **57** (bottom)
Sally Hill, Ultrasound Diagnostic Services **68**; **125** (bottom right)
The Hulton Picture Company **11**; **12**; **14**
The Hutchison Library **56** (bottom); **56** (top); **57** (center)
The Image Bank **2** (top left); **11**; **25**; **44**; **46** (bottom left); **79** (top left); **81**; **84** (center right); **117**; **134**
The Institute of Dermatology **74** (bottom); **90**; **132**
Dr. Helen Kingston **62** (bottom left); **62** (center left); **62** (top left); **62** (top right); **63**
Kunsthistorisches Museum, Vienna **43**
Living Technology **100** (bottom right)
Caroline Macy **15** (top)
Moorefield's Eye Hospital **101**
National Blood Transfusion Services **53**

National Medical Slide Bank, UK **51** (bottom); **62** (center right); **79** (bottom); **86** (center)
Pictor International Ltd **66** (center right); **77** (bottom right)
Planet Earth Pictures **10** (bottom left)
Science Photo Library **13** (bottom left); **13** (center right); **16**; **17**; **59**; **60**; **65**; **77** (bottom left); **77** (center right); **86** (bottom); **92**; **95**; **98**; **103**; **105**; **106**; **108**; **109**; **110** (bottom right); **110** (bottom left); **111**; **112**; **113**; **127**
S.D. Sigamoi, Christian Medical College Hospital, Vellore (*American Journal of Human Genetics*, 16. © 1964) **55**
St. Mary's Hospital Audio Visual Department **78** (center); **78** (bottom left); **80**; **87**; **91**; **100** (top right); **118** (bottom)
Tony Stone Worldwide **22**; **56** (center); **84** (center left); **116**
Sygma Ltd **46** (bottom left)
John Watney Photo Library **51** (top)
Werner Forman Archive **10** (top left)
Zefa **40**; **118** (top right)

Front cover photograph:
Nancy Brown/The Image Bank

Illustrators:
Russell Birkett
Sean Edwards
David Fathers
Bill le Fever
Evelina Frescura
Tony Graham
Andrew Green
Grundy & Northedge
Kevin Marks
Annabel Milne
Karin Murray

Commissioned photography:
Susannah Price
Steve Bartholemew

Airbrushing:
Paul Desmond
Roy Flooks

Index:
Sue Bosanko

The photograph of the St. Edward's crown on page **72** was reproduced with the kind permission of Her Majesty's Stationery Office.

Answers to IQ questions on page 20:
Chickens: You are looking at the left side of all the chickens except for the one in the bottom right corner.
Stick figures: The answer is E.